Synthesis Lectures on Digital Circuits & Systems

Series Editor

Mitchell A. Thornton, Southern Methodist University, Dallas, USA

This series includes titles of interest to students, professionals, and researchers in the area of design and analysis of digital circuits and systems. Each Lecture is self-contained and focuses on the background information required to understand the subject matter and practical case studies that illustrate applications. The format of a Lecture is structured such that each will be devoted to a specific topic in digital circuits and systems rather than a larger overview of several topics such as that found in a comprehensive handbook. The Lectures cover both well-established areas as well as newly developed or emerging material in digital circuits and systems design and analysis.

Tsutomu Sasao

Classification Functions for Machine Learning and Data Mining

 Springer

Tsutomu Sasao
Department of Computer Science
Meiji University
Kawasaki, Japan

ISSN 1932-3166   ISSN 1932-3174 (electronic)
Synthesis Lectures on Digital Circuits & Systems
ISBN 978-3-031-35346-8   ISBN 978-3-031-35347-5 (eBook)
https://doi.org/10.1007/978-3-031-35347-5

This Springer imprint is published by the registered company Springer Nature Switzerland AG
The registered company address is: Gewerbestrasse 11, 6330 Cham, Switzerland

Preface

A classification function is an integer-valued function that is defined by a training set. For instance, if we have a training set consisting of images of airplanes and automobiles, we can infer a function that can distinguish images of unseen airplanes and automobiles. Neural networks are often used for this purpose, but they require a long training time, a significant amount of hardware, and consume much power.

This book proposes a different approach to achieve the same goal. We use logic synthesis techniques and look-up tables (LUTs). Specifically, we reduce the number of variables and simplify sum-of-products expressions (SOPs), which makes training and hardware simpler. Additionally, this approach reduces power dissipation, although the accuracy is lower than that of neural networks. Applications of this approach include ultra-high-speed classifications, such as packet classification, network intrusion detection, and exotic particle detection in high-energy physics. It can also be used in medical diagnosis, where the size of the training set is small, and the data is imbalanced.

This book is suitable for graduate students and researchers in the fields of logic synthesis, machine learning, and data mining. Basic knowledge of logic synthesis is required to read this book, and knowledge of linear algebra and statistics is helpful. Most chapters contain many examples and exercises. And the appendix includes solutions.

Kawasaki, Japan   Tsutomu Sasao

Acknowledgements

This research is supported in part by the Grants in Aid for Scientific Research of JSPS. Many people were involved in this project: Jon T. Butler, Yukihiro Iguchi, Alan Mishchenko, Hiroki Nakahara, and Shinobu Nagayama.

Most materials in this book have been presented at various conferences: International Workshop on Logic and Synthesis (IWLS), International Symposium on Multiple-Valued Logic (ISMVL), International Conference on Computer-Aided Design (ICCAD), Asia South-Pacific Design Automation Conference (ASPDAC), and Workshop on Synthesis And System Integration of Mixed Information technologies (SASIMI), as well as journals: The Institute of Electronics, Information and Communication Engineers (IEICE), and Journal of Multiple-Valued Logic and Soft Computing. In many cases, reviewers' comments considerably improved the quality of the materials.

Numerous ideas were inspired by the students of Meiji University: Yuji Urano, Ichido Fumishi, Kyu Matsuura, and Kazuyuki Kai.

Prof. Jon T. Butler read through the entire manuscript repeatedly and made important corrections and improvements. Dr. Alan Mishchenko sent me a paper [1], which motivated me to write this book.

Reference

1. Chatterjee S (2018) Learning and memorization. International conference on machine learning (ICML 2018), Stockholm, Sweden, July 10-15, pp 754-762

Contents

About the Author

Tsutomu Sasao received B.E., M.E., and Ph.D. degrees in Electronics Engineering from Osaka University, Osaka Japan, in 1972, 1974, and 1977, respectively. He has held faculty/research positions at Osaka University, Japan; IBM T. J. Watson Research Center, Yorktown Height, NY; the Naval Postgraduate School, Monterey, CA; Kyushu Institute of Technology, Iizuka, Japan; and Meiji University, Kawasaki, Japan. His research areas include logic design and switching theory, representations of logic functions, and multivalued logic. He has published more than 10 books on logic design including, *Logic Synthesis and Optimization* (1993), *Representation of Discrete Functions* (1996), *Switching Theory for Logic Synthesis* (1999), *Logic Synthesis and Verification* (2002), *Progress in Applications of Boolean Functions* (2010), *Memory-Based Logic Synthesis* (2011), *Applications of Zero-Suppressed Decision Diagrams* (2015), and *Index Generation Functions* (2020). He has served Program Chairman for the IEEE International Symposium on Multiple-Valued Logic (ISMVL) many times. Also, he was the Symposium Chairman of the 28th ISMVL held in Fukuoka, Japan in 1998. He received the NIWA Memorial Award in 1979, Takeda Techno-Entrepreneurship Award in 2001, and Distinctive Contribution Awards from IEEE Computer Society MVL-TC for papers presented at ISMVLs in 1986, 1996, 2003, 2004, 2013, and 2019. He has served an associate editor of the *IEEE Transactions on Computers*. He is a Life Fellow of the IEEE.

Introduction

1

This chapter contains the aim and organization of the book.

1.1 Aim of this Book

This book shows a method to derive a simple **model** (e.g., classifier) that is consistent with a given set of examples (i.e., training set).

As a model, we assume a **sum-of-products expression** (**SOP**). The complexity is measured by the number of variables and the number of products in an SOP. When the model is implemented by hardware, **look-up tables** (**LUTs**) can be used.

Conventional machine learning systems are often implemented by neural networks, and learning is done by adjusting the weights of neurons. On the other hand, in the present method, learning is done by reduction of variables and simplifying SOPs. When the classifier is implemented by LUTs, the power dissipation can be reduced drastically. When the classifier is implemented by software, the set of rules are simple and easily interpretable. Unfortunately, the accuracy is not so good as classifiers using neural networks.

To read this book, a basic knowledge of logic design and discrete mathematics is necessary.

1.2 Organization of the Book

This book consists of 12 chapters. Figure 1.1 shows the relation among the chapters, where the arrows show the order to read the chapters. For example, Chaps. 7 and 9 can be read after reading Chap. 3. Among the chapters, Chaps. 2–4, and 6 are essential, so everybody

© The Author(s), under exclusive license to Springer Nature Switzerland AG 2024   1
T. Sasao, *Classification Functions for Machine Learning and Data Mining*,
Synthesis Lectures on Digital Circuits & Systems,
https://doi.org/10.1007/978-3-031-35347-5_1

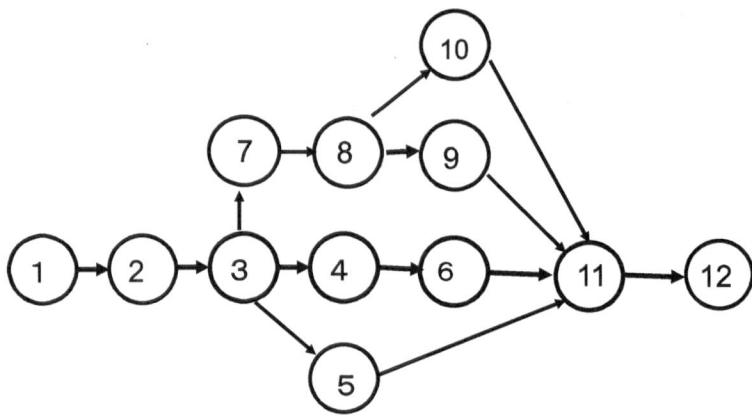

Fig. 1.1 Relation among chapters

is recommended to read these chapters. Sections 5.1 and 6.1 are advanced, so they can be skipped for the first reading.

Chapter 2 reviews basic properties of partially defined logic functions, and minimization of sum-of-products expressions.

Chapter 3 shows an exact method to minimize the number of variables.

Chapter 4 shows a heuristic method to reduce the number of variables. To select important variables, the **impurity measure** is introduced.

Chapter 5 derives the number of variables to represent random two-class functions. When n, the number of input variables, is sufficiently large, most random two-class functions can be represented with $p = \lceil \log_2(k_0 k_1) \rceil - 2$ variables, where k_1 and k_0 are the number of vectors in the ON and the OFF sets, respectively.

Chapter 6 introduces linear transformations of the input variables. With this technique, the number of variables can be reduced drastically.

Chapter 7 reviews terminology used in data mining and machine learning. It also shows the relation among logic design, machine learning, and data mining.

Chapter 8 shows methods to simplify SOPs with multi-valued inputs, and to minimize the number of variables. It also contains the experimental results for University of California-Irvine (UCI) data sets.

Chapter 9 shows two-class functions whose outcomes for unknown data are easy to predict by simplifying SOPs. This ability is the **generalization ability**.

Chapter 10 considers functions with continuous variables. First, it shows a method to convert continuous variables into discrete ones. Then, it shows the minimization results for UCI data sets.

Chapter 11 lists references on related works.

Chapter 12 summarizes the results obtained in the book, and shows remaining problems. At the end of chapters, exercises are shown. Exercises with (E) marks are relatively easy. The appendix contains solutions to the exercises.

Definitions and Basic Properties

2

This chapter defines two-class functions and classification functions. A two-class function classifies input vectors into two classes, while a classification function classifies input vectors into m classes, where $m > 2$.

2.1 Two-Class Function

Definition 2.1.1 Let ON, OFF, and DC be subsets of B^n, where $B = \{0, 1\}$, $ON \cap OFF = \varnothing$, $ON \cap DC = \varnothing$, $OFF \cap DC = \varnothing$, and $ON \cup OFF \cup DC = B^n$, where n shows the number of variables, and DC is the set of *don't cares*.

Consider a function f such that, for any $\vec{a} \in B^n$,

$$\vec{a} \in ON \Rightarrow f(\vec{a}) = 1,$$
$$\vec{a} \in OFF \Rightarrow f(\vec{a}) = 0.$$

When $DC = \varnothing$, f is **totally defined**, while when $DC \neq \varnothing$, f is **partially defined**.

For a totally defined function f, either the **ON set** or the **OFF set** can be used to specify f, since $OFF = B^n \setminus ON$.

For a partially defined function f, any two subsets of the three can be used to specify f. We assume that n is large, and the numbers of elements in ON and OFF are much smaller than that of DC. We call such functions **two-class functions**.

Example 2.1.1 Consider the partially defined function in Table 2.1, where the ON and the OFF sets are shown, but the vectors representing the DC set are omitted. Figure 2.1 is the

© The Author(s), under exclusive license to Springer Nature Switzerland AG 2024
T. Sasao, *Classification Functions for Machine Learning and Data Mining*,
Synthesis Lectures on Digital Circuits & Systems,
https://doi.org/10.1007/978-3-031-35347-5_2

Table 2.1 Partially defined function

		x_1	x_2	x_3	x_4	f
ON	$\vec{a}_1$	0	1	0	1	1
	$\vec{a}_2$	1	0	0	1	1
	$\vec{a}_3$	1	1	0	0	1
OFF	$\vec{b}_1$	0	0	0	1	0
	$\vec{b}_2$	0	1	1	0	0
	$\vec{b}_3$	1	0	0	0	0

Fig. 2.1 Four-variable partially
defined function

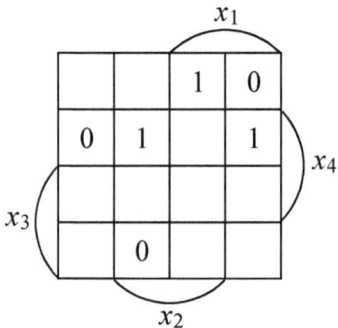

map of the function. In this case, the ON set consists of three vectors:

$$\{\vec{a}_1, \vec{a}_2, \vec{a}_3\},$$

while the OFF set consists of three vectors:

$$\{\vec{b}_1, \vec{b}_2, \vec{b}_3\}.$$

Figure 2.2 shows the map for the SOP with the fewest products:

$$\mathcal{F}_1 = x_1 x_4 \vee x_2 \bar{x}_3,$$

while Fig. 2.3 shows the SOP with the fewest variables:

$$\mathcal{F}_2 = x_1 x_4 \vee x_1 x_2 \vee x_2 x_4.$$

If the number of the products is the cost measure, then $\mathcal{F}_1$ is the optimum solution. However, if the number of the variables is the cost measure, then $\mathcal{F}_2$ is the optimum solution.   ∎

Fig. 2.2 SOP with the fewest products: $\mathcal{F}_1$

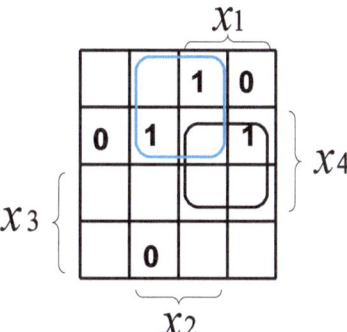

Fig. 2.3 SOP with the fewest variables: $\mathcal{F}_2$

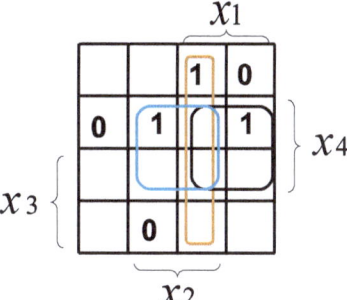

Definition 2.1.2 A **minimum SOP** is the SOP with the fewest products. Simplification of an SOP is to find an SOP with fewer productsthan the original SOP.

No efficient method to minimize literals is known [1]. Most SOP minimization algorithms first minimize the number of products, and then minimize the number of literals.

To find an SOP of a given function, we must cover all the **1-cells** by loops without covering any **0-cell**.

When the circuit is implemented by an AND-OR two-level circuit, the number of products is the cost measure, since it corresponds to the size of the circuit.

When the circuit is implemented by a memory or look-up table (LUT), the number of the variables is the cost measure, since it shows the size of the memory.

When the circuit is implemented by a circuit with fan-in limited gates, the total number of literals is the cost measure, since it corresponds to the area for the circuit.

For many partially defined functions, reducing the variables increases products, and vice versa. That it, there is a trade-off between the number of variables and the number of products in the SOP. So, we define another cost measure considering both the number of variables and products: the area of a programmable logic array (PLA).

Definition 2.1.3 The **PLA measure** for a two-class function is

$$n \cdot t,$$

where n is the number of variables, and t is the number of products in the SOP.

Example 2.1.2 The PLA measure for the SOP in Fig. 2.2 is $4 \times 2 = 8$. On the other hand, the PLA measure for the SOP in Fig. 2.3 is $3 \times 3 = 9$. ∎

2.2   Classification Function

Definition 2.2.1 A partially defined **classification function** f is a mapping $D \rightarrow M$, where $D \subset B^n$, $B = \{0, 1\}$, and $M = \{1, 2, \ldots, m\}$. Assume that D contains k distinct vectors; k is called the **weight of the function**. These vectors are **registered vectors**. For each registered vector, assign an integer between 1 and m, where $2 \leq m \leq k$. A **registered vector table** shows the **function value (class)** for each registered vector. A partially defined classification function f produces the corresponding function value when the input vector matches a registered vector. Let F_i be the set of registered vectors which map to $i \in M$, i.e., $F_i = f^{-1}(i)$. Then, $D = \bigcup_{i=1}^{m} F_i$. We assume that $F_i \neq \varnothing$ ($i = 1, 2, \ldots, m$).

Definition 2.2.2 An **index generation function** is a classification function, where k, the number of registered vectors, is equal to m, the number of the classes.

A classification function with $M = \{1, 2\}$ is the two-class function with function values 0 and 1 replaced by 1 and 2, respectively.

Example 2.2.1 The registered vector table in Table 2.2 shows a classification function with $n = 6$, $m = 2$ and $k = 8$. It is also a two-class function. ∎

A classification function $f : D \rightarrow M$ can be denoted by $(F_1, F_2, \ldots, F_m)$, if $F_i = f^{-1}(i)$ for all $i \in M$, and $M = \{1, 2, \ldots, m\}$.

For the next definition, two functions with different domains are considered. So, one has to be careful about determining m and, hence, M.

Definition 2.2.3 For **classification functions** $f : D \rightarrow M$ and $g : E \rightarrow M$ with the same range M=$\{1,2,\ldots,m\}$, where $2 \leq m \leq min\{|D|, |E|\}$, g is an **extension** of f if $f^{-1}(i) \subseteq g^{-1}(i)$ for all $i \in M$.

Table 2.2 Registered vector table

x_1	x_2	x_3	x_4	x_5	x_6	f
1	1	0	0	1	1	1
0	1	1	0	1	1	1
0	1	0	1	0	0	1
0	0	0	0	1	0	1
1	1	0	1	1	1	2
1	0	1	1	1	1	2
1	0	0	0	1	1	2
0	0	1	0	1	0	2

Example 2.2.2 Consider the case of $n = 4$. Let

$$F_1 = \{(0, 0, 0, 0), (0, 1, 0, 1), (0, 1, 1, 0)\}$$
$$F_2 = \{(0, 0, 1, 0), (1, 0, 0, 0), (1, 0, 1, 1)\}$$
$$E_1 = \{(0, 0, 0, 0), (0, 1, 0, 1), (0, 1, 1, 0), (1, 1, 0, 0)\}$$
$$E_2 = \{(0, 0, 1, 0), (1, 0, 0, 0), (1, 0, 1, 1), (1, 1, 0, 1)\}$$
$$G_1 = \{(0, 0, 0, 0), (0, 1, 0, 1), (0, 1, 1, 0), (1, 1, 0, 1)\}$$
$$G_2 = \{(0, 0, 1, 0), (1, 0, 0, 0), (1, 0, 1, 1), (1, 1, 0, 1)\}.$$

Then, (E_1, E_2) is an extension of (F_1, F_2), since $F_1 \subset E_1$, $F_2 \subset E_2$ and $E_1 \cap E_2 = \varnothing$. However, (G_1, G_2) is not an extension of (F_1, F_2), since $G_1 \cap G_2 \neq \varnothing$. ∎

Definition 2.2.4 For a subset $U \subseteq B^n$ and $S \subseteq \{1, 2, \ldots, n\}$, we denote by $U|_S$ the **projection** of U to S. In other words, $U|_S$ is the set of points obtained from U by considering only the jth components, where $j \in S$.

Example 2.2.3 Consider the case of $n = 4$. Let $U = \{(1, 0, 0, 1), (0, 1, 1, 0), (0, 0, 1, 0)\}$ and $S = \{2, 3\}$. Then, we have $U|_S = \{(*, 0, 0, *), (*, 1, 1, *), (*, 0, 1, *)\}$. ∎

Given a partially defined function, many extensions exist. In this paper, we seek the extension of f that depends on the fewest variables.

Definition 2.2.5 Let $F_i \subseteq B^n$ $(i = 1, 2, \ldots, m)$. Given a partially defined function $(F_1, F_2, \ldots, F_m)$, and a subset $S \subseteq \{1, 2, \ldots, n\}$, if $F_i|_S \cap F_j|_S = \varnothing$, $(i \neq j)$ holds, then S is a **support set**. In such a case, $(F_1|_S, F_2|_S, \ldots, F_m|_S)$ is independent of the variable x_j, $j \in \{1, 2, \ldots, n\} \setminus S$, and the variable x_j is **redundant**.

Table 2.3 Classification function with reduced variables

x_1	x_2	x_3	x_4	f	TAG
1	1	0	0	1	1
0	1	1	0	1	2
0	1	0	1	1	3
0	0	0	0	1	4
1	1	0	1	2	5
1	0	1	1	2	6
1	0	0	0	2	7
0	0	1	0	2	8

Example 2.2.4 Consider the function (F_1, F_2) shown in Table 2.2. In this case, $S = \{1, 2, 3, 4\}$ is a support set, since for

$$F_1|_S = \{(1, 1, 0, 0, *, *), (0, 1, 1, 0, *, *),$$
$$(0, 1, 0, 1, *, *), (0, 0, 0, 0, *, *)\},$$
$$F_2|_S = \{(1, 1, 0, 1, *, *), (1, 0, 1, 1, *, *),$$
$$(1, 0, 0, 0, *, *), (0, 0, 1, 0, *, *)\}$$

$F_1|_S \cap F_2|_S = \varnothing$ holds. Thus, this function can be represented by four variables as shown in Table 2.3.

However, $T = \{1, 2, 3\}$ is not a support set, since for

$$F_1|_T = \{(1, 1, 0, *, *, *), (0, 1, 1, *, *, *),$$
$$(0, 1, 0, *, *, *), (0, 0, 0, *, *, *)\},$$
$$F_2|_T = \{(1, 1, 0, *, *, *), (1, 0, 1, *, *, *),$$
$$(1, 0, 0, *, *, *), (0, 0, 1, *, *, *)\}$$

$F_1|_T \cap F_2|_T = \{(1, 1, 0, *, *, *)\} \neq \varnothing$ holds.   ■

2.3   Remarks

This chapter defined two-class functions and classification functions. It also shows the condition to reduce the number of variables.

This chapter is based on [2–6].

References

1. Muroga S (1979) Logic design and switching theory. Wiley-Interscience Publication
2. Ibaraki T (2011) Partially defined Boolean functions, Chapter 8. In: Crama Y, Hammer PL (eds) Boolean functions - theory, algorithms and applications. Cambridge University Press, New York
3. Sasao T (2011) Memory-based logic synthesis. Springer
4. Sasao T (2019) Index generation functions. Morgan & Claypool
5. Sasao T (2021) On a design of multi-layer LUT network. IWLS, Online, July 19–21
6. Sasao T (2023) Data mining using multi-valued logic minimization. ISMVL, May 22-24

Minimization of the Number of Variables: Exact Method

3

This chapter shows an exact method to minimize the number of variables in partially defined functions using difference vectors.

3.1 Definitions

Given a classification function $(F_1, F_2, \ldots, F_m)$, to minimize the number of variables, we seek the fewest variables that distinguish all F_i.

Definition 3.1.1 Let $f : D \to M$ be a classification function with n variables, where $D \subseteq B^n$, $B = \{0, 1\}$, and $M = \{1, 2, \ldots, m\}$. If there exist $\vec{a} = (a_1, a_2, \ldots, a_n)$, $\vec{b} = (b_1, b_2, \ldots, b_n) \in D$ and $i \in \{1, 2, \ldots, n\}$ such that $a_j = b_j$ for every $j \in \{1, 2, \ldots, n\} \setminus \{i\}$,[1] $a_i \neq b_i$ and $f(\vec{a}) \neq f(\vec{b})$, then f is said to **depend** on the i-th variable. In this case, the i-th variable is **essential**.

Example 3.1.1 Consider the function (F_1, F_2) shown in Table 3.1. This function depends on x_4. This can be verified from the fact

$$\vec{a}_1 = (1, 1, 0, 0, 1, 1) \in F_1 \quad \text{and}$$
$$\vec{b}_1 = (1, 1, 0, 1, 1, 1) \in F_2.$$

This function also depends on x_2. This can be verified from the fact

---

[1] The symbol $\setminus$ denotes set subtraction, i.e. $a \in A \setminus B$ iff $a \in A$ and $a \notin B$.

© The Author(s), under exclusive license to Springer Nature Switzerland AG 2024
T. Sasao, *Classification Functions for Machine Learning and Data Mining*,
Synthesis Lectures on Digital Circuits & Systems,
https://doi.org/10.1007/978-3-031-35347-5_3

Table 3.1 Explanation of support set

		x_1	x_2	x_3	x_4	x_5	x_6	f
F_1	$\vec{a}_1$	1	1	0	0	1	1	1
	$\vec{a}_2$	0	1	1	0	1	1	1
	$\vec{a}_3$	0	0	0	0	1	0	1
F_2	$\vec{b}_1$	1	1	0	1	1	1	2
	$\vec{b}_2$	1	0	0	0	1	1	2
	$\vec{b}_3$	0	1	0	1	0	0	2

$$\vec{a}_1 = (1, 1, 0, 0, 1, 1) \in F_1 \text{ and}$$
$$\vec{b}_2 = (1, 0, 0, 0, 1, 1) \in F_2.$$

∎

Definition 3.1.2 In a partially defined function $(F_1, F_2, \ldots, F_m)$, let $\vec{a} \in F_i$, and $\vec{b} \in F_j$, $(i \neq j)$. The vector $\vec{d} = \vec{a} \oplus \vec{b}$ is a **difference vector**, where $\oplus$ denotes the bitwise exor operator. The set of the difference vectors is denoted by D_f. The set of vectors that is obtained from D_f by deleting the vector $\vec{b}$ satisfying $\vec{a} < \vec{b}$ and $\vec{a} \in D_f$ is the set of the **minimal difference vectors**, and denoted by MD_f.

Theorem 3.1.1 *The necessary and sufficient condition that the variable x_i is essential in the function f is the set of difference vectors contains the unit vector $\vec{e}_i$ (the vector where only the i-th component is one, and others are zero).*

Theorem 3.1.2 *To represent a partially defined classification function $(F_1, F_2, \ldots, F_m)$, at least $\lceil \log_2 m \rceil$ variables are necessary.*

3.2 Exact Algorithm

The following algorithm is an extension of the one for two-class function [1, 2].

Algorithm 3.2.1 (Minimization of the Number of Variables)

1. Determine the set of the essential variables.
2. Obtain the set of the minimal difference vectors MD_f of a partially defined function $(F_1, F_2, \ldots, F_m)$.
3. For each $\vec{d} = (d_1, d_2, \ldots, d_n) \in MD_f$, realize a clause

$$C(\vec{d}) = z_1 \vee z_2 \vee \cdots \vee z_n,$$

where

$$z_j = \begin{cases} y_i & \text{(when } d_j = 1) \\ 0 & \text{(when } d_j = 0). \end{cases}$$

4. Realize a product-of-sums expression:

$$R = \bigwedge_{\vec{d} \in DF} C(\vec{d}).$$

5. Convert R into an SOP [3], and simplify it. A product term with the fewest literals shows a **minimum support set**.

Example 3.2.2 Consider the decision function (F_1, F_2) shown in Table 3.1.

1. Since
   $\vec{a}_1 \oplus \vec{b}_1 = (0, 0, 0, 1, 0, 0)$, and $\vec{a}_1 \oplus \vec{b}_2 = (0, 1, 0, 0, 0, 0)$, x_2 and x_4 are essential.
2. We can generate the set of difference vectors for (F_1, F_2) as follows:
   For $(x_2, x_4) = (0, 0)$ the set of difference vector is $(\vec{a}_3, \vec{b}_2) = (1, 0, 0, 0, 0, 1)$.
   For $(x_2, x_4) = (0, 1), (1, 0)$ and $(1, 1)$, the set of difference vectors is null.
3. The set of the minimal difference vectors is

$$MD_f = \{(0, 0, 0, 1, 0, 0), (0, 1, 0, 0, 0, 0), (1, 0, 0, 0, 0, 1)\}.$$

4. The set of clauses derived from the minimal difference vectors is:

$$C(\vec{a}_1, \vec{b}_1) = y_4,$$
$$C(\vec{a}_1, \vec{b}_2) = y_2, \text{ and}$$
$$C(\vec{a}_3, \vec{b}_2) = y_1 \vee y_6.$$

5. By obtaining the product of all the clauses, we have

$$R = y_2 y_4 (y_1 \vee y_6).$$

6. By converting this expression into an SOP, and by simplifying, we have
   $R = y_1 y_2 y_4 \vee y_2 y_4 y_6$.
7. The minimum support sets are $\{1, 2, 4\}$ and $\{2, 4, 6\}$. ∎

3.3 Remarks

This chapter showed a method to minimize the number of variables for partially defined functions. It involves a **minimum covering problem**, which is very time-consuming. This method is practical for functions with up to 100 variables. Experimental results are shown in Sects. 5.2, 6.3, and 6.4.2. Note that minimization of the number of variables is easier than that of SOPs (See Exercise 3.5). This chapter is based on [4, 5].

3.4   Exercises

3.1 *(E) Consider the function shown in Table 2.2. Find all the minimum sets of variables to represent the function.*

3.2 *(E) Let D_f be the set of difference vectors of a classification function $f(x_1, x_2, \ldots, x_n)$. Show that if $\vec{e}_1 = (1, 0, \ldots, 0, 0) \notin D_r$, then f can be represented without x_1.*

3.3 *(E) Let D_f be the set of difference vectors of a classification function $f(x_1, x_2, \ldots, x_n)$. Show that if $\vec{e}_1 = (1, 0, \ldots, 0, 0) \in D_r$, then x_1 is essential.*

3.4 *Let D_f be the set of difference vectors of a classification function $f(x_1, x_2, \ldots, x_n)$. Show that if $\vec{e}_1 = (1, 0, 0, \ldots, 0, 0) \notin D_f$ and $\vec{e}_2 = (0, 1, 0, \ldots, 0, 0) \notin D_f$ and $(\vec{e}_1 \vee \vec{e}_2) \notin D_f$, then f can be represented without $\{x_1, x_2\}$.*

3.5 *Show that minimization of the number of variables is easier than that of the number of products in an SOP.*

3.6 *(E) Minimize the number of variables for the function in Table 3.2.*

3.7 *(E) Consider the function f_5 in Table 3.3. Show that f_5 can be represented by either $\{x_1, x_2, x_3, x_4\}$ or y_4.*

3.8 *Consider the partially defined function F, where the ON set and the OFF set are represented by h_{ON} and h_{OFF}, respectively, and*

Table 3.2 Classification function

		x_1	x_2	x_3	x_4	f
F_1	$\vec{a}_1$	1	0	0	0	1
	$\vec{a}_2$	0	1	0	0	1
F_2	$\vec{b}_1$	0	1	1	0	2
	$\vec{b}_2$	1	1	0	1	2

Table 3.3 Example function

x_1	x_2	x_3	x_4	y_4	f_5
0	0	0	0	0	0
1	0	0	0	1	1
1	1	0	0	0	0
1	1	1	0	1	1
1	1	1	1	0	0

$$h_{ON} = (x_1 \oplus x_2 \oplus \cdots \oplus x_n)yz \vee \bar{x}_1\bar{x}_2 \cdots \bar{x}_n \bar{y}z,$$
$$h_{OFF} = (x_1 \oplus x_2 \oplus \cdots \oplus x_n)\bar{y}z \vee x_1 x_2 \cdots x_n y\bar{z}.$$

Show the following:
If n is even, then F can be represented by an SOP with only two products:

$$\mathcal{F}_1 = yz \vee \bar{x}_1\bar{x}_2 \cdots \bar{x}_n.$$

Also, F can be represented by the SOP without z, but in this case, the SOP requires $n + 1$ products:

$$\mathcal{F}_2 = \bar{x}_1 y \vee \bar{x}_2 y \vee \cdots \vee \bar{x}_n y \vee \bar{x}_1\bar{x}_2 \cdots \bar{x}_n.$$

3.9 *Let MD_f be the set of minimal difference vectors for an n-variable classification function. Show that the number of elements in MD_f is at most*

$$\binom{2^n}{\lceil \frac{n}{2} \rceil}.$$

References

1. Sasao T (2008) On the number of variables to represent sparse logic functions, ICCAD-2008. California, USA, San Jose, pp 45–51
2. Kuntzmann J (1967) Algèbre de Boole, Dunod, Paris, 1965. Fundamental Boolean Algebra, Blackie and Son Limited, London and Glasgow, English translation
3. Sasao T (1999) Switching theory for logic synthesis. Kluwer Academic Publishers
4. Sasao T (2019) On a minimization of variables to represent sparse multi-valued input decision functions. ISMVL, Fredericton, Canada, pp 182–187
5. Sasao T (2021) On the number of variables to represent classification functions using linear decompositions. SASIMI, March 29–30. (Virtual workshop)

Minimization of the Number of Variables: Heuristic Method

4

This chapter shows a method to select important variables to represent a classification function. To do it, the impurity measure is introduced.

4.1 Impurity Measure

In this part, we introduce the impurity measure [1].

Definition 4.1.1 Let $f(x_1, x_2, \ldots, x_n)$ be a partially defined classification function to represent. Let $\vec{a} = (a_1, a_2, \ldots, a_n)$ be a binary vector showing the selected variables. $a_i = 1$ shows that x_i is selected for expansion. Let $Size(j)$ be the number of vectors that belong to the partition j generated by $\vec{a}$. Let $Hist(j, Value)$ be the number of vectors that belong to the partition j and whose value corresponds to $Value$. In this case, the following relations hold:

$$\sum_{Value=1}^{m} Hist(j, Value) = Size(j), \quad \sum_{j=0}^{2^t-1} Size(j) = k.$$

k denotes the total number of registered vectors. t denotes the number of variables such that $a_i = 1$. $\vec{a}$ represents the distribution of variables such that $\sum_{i=1}^{n} a_i = t$. Thus, in this case, the **impurity measure** of the function is

$$\mu(\vec{a}) = \sum_{j=0}^{2^t-1} \left[Size(j)^2 - \sum_{Value=1}^{m} Hist(j, Value)^2 \right] \tag{4.1}$$

T. Sasao, *Classification Functions for Machine Learning and Data Mining*,
Synthesis Lectures on Digital Circuits & Systems,
https://doi.org/10.1007/978-3-031-35347-5_4

When $\mu(\vec{a}) = 0$, the classification function f can be represented by the variables specified by $\vec{a}$.

Example 4.1.1 To illustrate the impurity measure, we use Figs. 4.1, 4.2, 4.3, 4.4 and 4.5.

- In Fig. 4.1, there are 13 blue circles. It consists of single kind of items, so it is pure. Since, $k = k_1 = 13$, we have $\mu = 13^2 - 13^2 = 0$. Thus, the impurity measure is 0.

Fig. 4.1 Pure

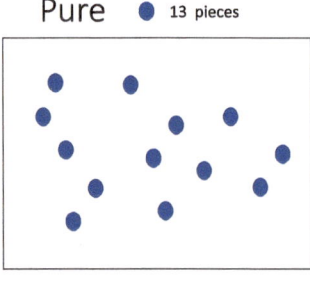

Fig. 4.2 Slightly impure

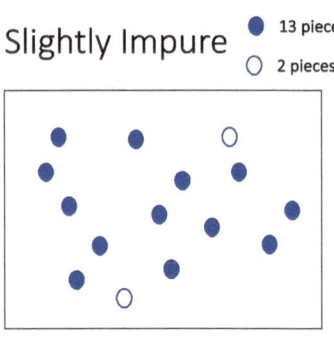

Fig. 4.3 Impure

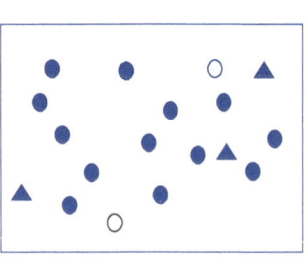

Fig. 4.4 Very impure

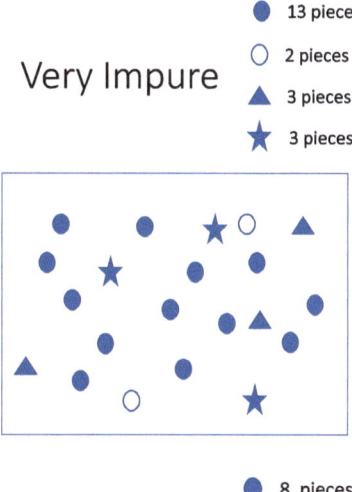

Fig. 4.5 Mixed

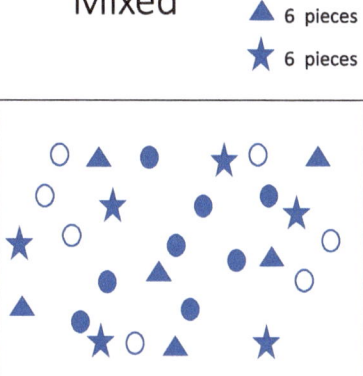

- In Fig. 4.2, two circles without color are added. So, it is slightly impure. Since, $k = 15$, $k_1 = 13$, $k_2 = 2$, we have $\mu = 15^2 - (13^2 + 2^2) = 52$. Thus, the impurity measure is 52.
- In Fig. 4.3, three blue triangles were added. So, it is impure. Since, $k = 18$, $k_1 = 13$, $k_2 = 2$, $k_2 = 3$, we have $\mu = 18^2 - (13^2 + 2^2 + 3^2) = 142$. Thus, the impurity measure is 142.
- In Fig. 4.4, three blue stars were added. So, it is very impure. Since, $k = 21$, $k_1 = 13$, $k_2 = 2$, $k_2 = 3$, $k_3 = 3$, we have $\mu = 18^2 - (13^2 + 2^2 + 3^2 + 3^2) = 250$. Thus, the impurity measure is 250.
- In Fig. 4.5, five blue circles were changed to circles without color. Three blue stars were added, and three blue triangles were added. So, it is mixed. Since, $k = 27$, $k_1 = 8$, $k_2 = $

7, $k_2 = 6$, $k_3 = 6$, we have $\mu = 27^2 - (8^2 + 7^2 + 6^2 + 6^2) = 544$. Thus, the impurity measure is 544.

In this way, the impurity can be derived.   ∎

4.2   Heuristic Algorithm

Algorithm 4.2.1 (*A heuristic method to reduce the number of variables*) Given a registered vector table of a classification function f.

1. Compute the impurity measure $\mu(\vec{e}_i)$ for $i = 1, 2, \ldots, n$. Note that $\vec{e}_i$ denotes the unit vector, where only the i-th element is 1, and other elements are 0s.
2. Assume that $\mu(\vec{e}_i)$ is the minimum. Let $\vec{a} \leftarrow e_i$. $\vec{a}$ shows the set of selected variables.
3. Select a variable x_j from the remaining set of variables, so that the measure is minimized for the resulting decision tree. Let $\vec{a} \leftarrow \vec{a} \vee \vec{e}_j$.
4. If $\mu(\vec{a}) > 0$, then go to step 3, else stop.

Since this algorithm is a heuristic one, it does not always produce a minimum solution.

Example 4.2.1 Consider the 6-variable function shown in Table 4.1. When the function is expanded by x_1, we have $\vec{a} = (1, 0, 0, 0, 0, 0)$ and Table 4.2. There are two blocks.
  Note that $Size(0) = 5$ and $Size(1) = 4$.
  For the partition, $x_1 = 0$,
$Hist(0, 1) = 2, Hist(0, 2) = 1, Hist(0, 3) = 2.$
  For the partition $x_1 = 1$,
$Hist(1, 1) = 1, Hist(1, 2) = 2, Hist(1, 3) = 1.$

Table 4.1 6-variable partially defined classification function

x_1	x_2	x_3	x_4	x_5	x_6	f
1	1	0	0	1	1	3
0	1	1	0	1	1	3
0	0	0	0	1	0	3
1	1	0	1	1	1	2
1	0	0	0	1	1	2
0	1	0	1	0	0	2
1	0	1	1	1	1	1
0	0	1	0	1	0	1
0	0	0	1	0	1	1

Table 4.2 When the function is expanded by x_1

x_1	x_2	x_3	x_4	x_5	x_6	f
0	1	1	0	1	1	3
0	0	0	0	1	0	3
0	1	0	1	0	0	2
0	0	1	0	1	0	1
0	0	0	1	0	1	1
1	1	0	0	1	1	3
1	1	0	1	1	1	2
1	0	0	0	1	1	2
1	0	1	1	1	1	1

Table 4.3 When the function is expanded by x_2

x_1	x_2	x_3	x_4	x_5	x_6	f
0	**0**	0	0	1	0	3
1	**0**	0	0	1	1	2
0	**0**	1	0	1	0	1
1	**0**	1	1	1	1	1
0	**0**	0	1	0	1	1
1	**1**	0	0	1	1	3
0	**1**	1	0	1	1	3
1	**1**	0	1	1	1	2
0	**1**	0	1	0	0	2

Thus, the impurity measure is

$$\mu(\vec{a}) = [5^2 - (2^2 + 1^2 + 2^2)] + [4^2 - (1^2 + 2^2 + 1^2)]$$
$$= 41 - 15 = 26.$$

When the function is expanded by x_2, we have $\vec{a} = (0, 1, 0, 0, 0, 0)$ and Table 4.3. Note that $Size(0) = 5$ and $Size(1) = 4$.

For the partition $x_2 = 0$,
$Hist(0, 1) = 3, Hist(0, 2) = 1, Hist(0, 3) = 1$.

For the partition $x_2 = 1$,
$Hist(1, 1) = 0, Hist(1, 2) = 2, Hist(1, 3) = 2$.

Thus, the impurity measure is

$$\mu(\vec{a}) = [5^2 - (3^2 + 1^2 + 1^2)] + [4^2 - (0^2 + 2^2 + 2^2)]$$
$$= 41 - 19 = 22.$$

Table 4.4 When the function is expanded by x_5

x_1	x_2	x_3	x_4	x_5	x_6	f
0	1	0	1	0	0	2
0	0	0	1	0	1	1
1	1	0	0	1	1	3
0	1	1	0	1	1	3
0	0	0	0	1	0	3
1	1	0	1	1	1	2
1	0	0	0	1	1	2
1	0	1	1	1	1	1
0	0	1	0	1	0	1

When the function is expanded by x_5, we have $\vec{a} = (0, 0, 0, 0, 1, 0)$ and Table 4.4. Note that $Size(0) = 2$ and $Size(1) = 7$.

For the partition $x_5 = 0$,
$Hist(0, 1) = 1, Hist(0, 2) = 1, Hist(0, 3) = 0.$

For the partition $x_5 = 1$,
$Hist(1, 1) = 2, Hist(1, 2) = 2, Hist(1, 3) = 3.$

Thus, the measure is

$$\mu(\vec{a}) = [2^2 - (1^2 + 1^2 + 0^2)] + [7^2 - (2^2 + 2^2 + 3^2)]$$
$$= 53 - 19 = 34.$$

In this way, for each variable, we compute the impurity measure. In summary, when the function is expanded by x_i, the measures are

$$x_1 : \mu = 26, \ x_2 : \mu = 22.$$
$$x_3 : \mu = 26, \ x_4 : \mu = 22.$$
$$x_5 : \mu = 34, \ x_6 : \mu = 30.$$

Since x_2 yields the smallest measures, we select x_2 to expand the function.[1]

In the next step, we select the second variable in a similar way, and find that when the function is expanded with x_2 and x_4, the measure becomes minimum. In this case, we have $\vec{a} = (0, 1, 0, 1, 0, 0)$ and Table 4.5 shows the partition. There are four blocks.

Note that $Size(00) = 3, Size(01) = 2, Size(10) = 2, Size(11) = 2.$

For the partition $(x_2, x_4) = 00$,
$Hist(00, 1) = 1, Hist(00, 2) = 1, Hist(00, 3) = 1.$

---

[1] If we select x_4, then in the next step, we have to select x_2, yielding the same solution.

Table 4.5 When the function is expanded by x_2 and x_4

x_1	x_2	x_3	x_4	x_5	x_6	f
0	0	0	0	1	0	3
1	0	0	0	1	1	2
0	0	1	0	1	0	1
0	0	0	1	0	1	1
1	0	1	1	1	1	1
1	1	0	0	1	1	3
0	1	1	0	1	1	3
1	1	0	1	1	1	2
0	1	0	1	0	0	2

Table 4.6 When the function is expanded by x_1, x_2 and x_4

x_1	x_2	x_3	x_4	x_5	x_6	f
0	0	0	0	1	0	3
0	0	1	0	1	0	1
0	0	0	1	0	1	1
0	1	1	0	1	1	3
0	1	0	1	0	0	2
1	0	0	0	1	1	2
1	0	1	1	1	1	1
1	1	0	0	1	1	3
1	1	0	1	1	1	2

For the partition $(x_2, x_4) = 01$,
$Hist(00, 1) = 2, Hist(00, 2) = 0, Hist(00, 3) = 0$.
For the partition $(x_2, x_4) = 10$,
$Hist(00, 1) = 0, Hist(00, 2) = 0, Hist(00, 3) = 2$.
For the partition $(x_2, x_4) = 11$,
$Hist(00, 1) = 0, Hist(00, 2) = 2, Hist(00, 3) = 0$.
Thus, the measure is

$$\mu(\vec{a}) = [3^2 - (1^2 + 1^2 + 1^2)] + [2^2 - (2^2 + 0^2 + 0^2)] +$$
$$[2^2 - (0^2 + 0^2 + 2^2)] + [2^2 - (0^2 + 2^2 + 0^2)] = 6.$$

In the next step, we select the third variable in a similar way, and find that when the function is expanded with x_1, x_2 and x_4, the measure becomes minimum. In this case, we have $\vec{a} = (1, 1, 0, 1, 0, 0)$, and Table 4.6 shows the partition. There are eight blocks.

Note that

$$Size(000) = 2, \quad Size(001) = 1, \quad Size(010) = 1,$$
$$Size(011) = 1, \quad Size(100) = 1, \quad Size(101) = 1,$$
$$Size(110) = 1, \quad Size(111) = 1.$$

For the partition $(x_1, x_2, x_4) = 000$,
$Hist(000, 1) = 1, Hist(000, 2) = 0, Hist(000, 3) = 1$.
For the partition $(x_1, x_2, x_4) = 001$,
$Hist(001, 1) = 1, Hist(001, 2) = 0, Hist(001, 3) = 0$.
For the partition $(x_1, x_2, x_4) = 010$,
$Hist(010, 1) = 0, Hist(010, 2) = 0, Hist(010, 3) = 1$.
For the partition $(x_1, x_2, x_4) = 011$,
$Hist(011, 1) = 0, Hist(011, 2) = 1, Hist(011, 3) = 0$.
For the partition $(x_1, x_2, x_4) = 100$,
$Hist(100, 1) = 0, Hist(100, 2) = 1, Hist(100, 3) = 0$.
For the partition $(x_1, x_2, x_4) = 101$,
$Hist(101, 1) = 1, Hist(101, 2) = 0, Hist(101, 3) = 0$.
For the partition $(x_1, x_2, x_4) = 110$,
$Hist(110, 1) = 0, Hist(110, 2) = 0, Hist(110, 3) = 1$.
For the partition $(x_1, x_2, x_4) = 111$,
$Hist(111, 1) = 0, Hist(111, 2) = 1, Hist(111, 3) = 0$.
Thus, the measure is

$$
\begin{aligned}
\mu(\vec{a}) = {} & [2^2 - (1^2 + 0^2 + 1^2)] + [1^2 - (1^2 + 0^2 + 0^2)] + \\
& [1^2 - (0^2 + 0^2 + 1^2)] + [1^2 - (0^2 + 1^2 + 0^2)] + \\
& [1^2 - (0^2 + 1^2 + 0^2)] + [1^2 - (1^2 + 0^2 + 0^2)] + \\
& [1^2 - (0^2 + 0^2 + 1^2)] + [1^2 - (0^2 + 1^2 + 0^2)] = 2.
\end{aligned}
$$

And, finally, we select the fourth variable x_3. In this case, we have $\vec{a} = (1, 1, 1, 1, 0, 0)$ and $\mu(\vec{a}) = 0$. Thus, the function is represented by (x_1, x_2, x_3, x_4).

Figure 4.6 shows the decision tree for the function. For each stage, we selected the variable that minimizes the impurity measure. Each vector $\vec{a}$ shows the selected variables. Each leaf node is a unique integer or an empty node. Note that in Eq. (4.1), the first $\sum$ sums for different assignments of variables, while the second $\sum$ sums for different values of variables.   ∎

For most sparse random functions, the numbers of variables can be reduced, which is shown in Chap. 6. However, for some functions, no variable can be removed.

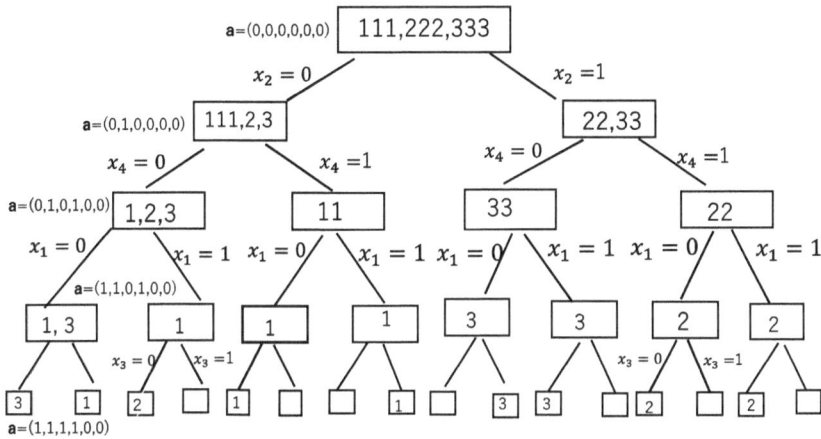

Fig. 4.6 Decision tree for the function in Table 4.1

Table 4.7 Function with all essential variables

x_1	x_2	x_3	x_4	x_5	x_6	f
0	0	0	0	0	0	0
1	0	0	0	0	0	1
0	1	0	0	0	0	1
0	0	1	0	0	0	1
0	0	0	1	0	0	1
0	0	0	0	1	0	1
0	0	0	0	0	1	1

Example 4.2.2 Consider the function in Table 4.7. Note that all the variables are essential. So, no variable can be eliminated at all. ∎

Even for such functions, the number of variables can be reduced by using linear transformations, which is shown in Chap. 5.

4.3   Remarks

This chapter showed a heuristic method to reduce the number of variables for classification functions. The method using the impurity measure is practical for functions with more than a thousand variables. Experimental results are shown in Chaps. 5 and 6. This chapter is based on [1]. The idea of the impurity measure is similar to the **Gini index** [2]. Most methods try to reduce the number of nodes in decision trees, but the method in this section tries to reduce the number of variables for expansion.

4.4   Exercises

4.1 *(E) Let k be the number of registered vectors, and n be the number of variables in the registered vector table. Estimate the size of the memory necessary to reduce the number of variables by Algorithm 4.2.1.*

4.2 *Derive the impurity measure of the function in Table 3.1 using Definition 4.1.1.*

References

1. Sasao T (2020) Reduction methods of variables for large-scale classification functions. In: IWLS, July 2020, pp 82–87
2. Breiman L, Friedman JH, Stone CJ, Olshen RA (1984) Classification and regression trees. CRC Press, New York

Two-Class Functions

5

This chapter shows that when n is sufficiently large, most random two-class functions can be represented with $p = \lceil \log_2(k_0 k_1) \rceil - 2$ variables, where k_1 and k_0 are the numbers of vectors in the ON and the OFF sets, respectively. It also shows experimental results for random and non-random two-class functions.

5.1 Statistical Analysis

To analyze random functions, we need two definitions:

Definition 5.1.1 ([1]) Let $f(X)$ be a logic function, and (X_1, X_2) be a partition of the input variables, where $X_1 = (x_1, x_2, \ldots, x_k)$ and $X_2 = (x_{k+1}, x_{k+2}, \ldots, x_n)$. The **decomposition chart** for f is a two-dimensional matrix with 2^k columns and 2^{n-k} rows, where each column and row is labeled by a unique binary code, and each element corresponds to the truth value of f.

Definition 5.1.2 ([2]) In probability theory and statistics, a collection of random variables is **independent and identically distributed** (**IID**) if each random variable has the same probability distribution as the others and all are mutually independent.[1]

---

[1] Suppose A, B, C are three events. If $P(AB) = P(A)P(B)$, $P(BC) = P(B)P(C)$, $P(AC) = P(A)P(C)$, $P(ABC) = P(A)P(B)P(C)$ are satisfied, then the events A, B, C are mutually independent, where $P(A)$ is the probability that A occurs, $P(B)$ is the probability that B occurs, and $P(C)$ is the probability that C occurs.

© The Author(s), under exclusive license to Springer Nature Switzerland AG 2024
T. Sasao, *Classification Functions for Machine Learning and Data Mining*,
Synthesis Lectures on Digital Circuits & Systems,
https://doi.org/10.1007/978-3-031-35347-5_5

IID is often used in machine learning [3]. In this part, we assume that the values of the truth table of the function are IID. A truth table of an n-variable function with the IID property can be generated as follows:

Algorithm 5.1.1 (Algorithm to generate a truth table with the IID property)
Assume that there are 2^n balls in a box. Among them, k_0 balls are white (0), k_1 balls are black (1), and the remaining balls are red (*don't care*). Consider a truth table of 2^n elements. Set the values of all the elements to *don't care*. Let the index of the truth table be 0.

1. Take a ball randomly from the box.
2. Write a 1 (0) into the index position of the truth table if the ball is black (white).
3. Then, return the ball into the box.
4. Increment the index of the truth table by one.
5. Repeat above steps 2^n times to make a truth table of 2^n elements.

From the **law of large numbers**, we have:

Lemma 5.1.1 *Let f be a function generated by Algorithm 5.1.1. The expected number of 1's and 0's in the truth table for f is k_1, and k_0, respectively.*

Lemma 5.1.2 *Consider a set of random functions, where k_0 combinations are mapped to 0, k_1 combinations are mapped to 1, and the other $2^n - (k_0 + k_1)$ combinations are undefined.[2] Let η be the probability that f can be represented by using only $x_1, x_2, \ldots, x_{p-1}$, and x_p, where $p < n$. Then,*

$$\eta \geq (1 - \tilde{\alpha}_0)^{k_1},$$

where $\tilde{\alpha}_0 = \frac{k_0}{2^p}$.

(Proof) Let $f(X_1, X_2)$ be a partially defined function, where $X_1 = (x_1, x_2, \ldots, x_p)$ and $X_2 = (x_{p+1}, x_{p+2}, \ldots, x_n)$. Consider the decomposition chart of $f(X_1, X_2)$, where X_1 labels the columns, and X_2 labels the rows. If no column has both 0 and 1, a totally defined function can be formed by setting all column entries to the same value, yielding a function independent of X_2. From here, we obtain the probability η.

Assume that k_0 0's are already distributed to the decomposition chart. Thus, at most k_0 columns have 0's. Next, we distribute k_1 1's to the decomposition chart. The probability of distributing a single 1 to a column not containing 0's is at least $\frac{2^p - k_0}{2^p} = 1 - \tilde{\alpha}_0$. Thus,

---

[2] This set of random functions, and the set of functions generated by Algorithm 5.1.1, are different. However, to make the analysis simple, we assume that the set of the random functions has the IID property. In Lemma 5.1.5, we show that for most functions generated by Algorithm 5.1.1, the number of true and false minterms are near k_1 and k_0, respectively.

Fig. 5.1 Decomposition chart for $f(X_1, X_2)$

			0	0	0	0	1	1	1	1	x_1
			0	0	1	1	0	0	1	1	x_2
			0	1	0	1	0	1	0	1	x_3
0	0	0	0							
0	0	1					1	1		0
0	1	0								
0	1	1								
1	0	0					1	0		
1	0	1								
1	1	0								
1	1	1						1	0	
x_4	x_5	x_6								

the probability of distributing k_1 1's to the columns without 0's is larger than or equal to $(1 - \tilde{\alpha}_0)^{k_1}$, by the IID assumption. Hence, we have the relation: $\eta \geq (1 - \tilde{\alpha}_0)^{k_1}$. $\square$

Example 5.1.1 Figure 5.1 shows an example of a decomposition chart, where $X_1 = (x_1, x_2, x_3)$ labels columns, and $X_2 = (x_4, x_5, x_6)$ labels rows. It has $k_0 = 4$ zeros, and $k_1 = 4$ ones. The blank entries are *don't cares*. In this chart, four columns have 0's, and $\tilde{\alpha} = \frac{k_0}{2^p} = \frac{4}{2^3} = \frac{1}{2}$. The probability of distributing a single 1 to a column not containing 0's is at least $\frac{2^p - k_0}{2^p} = \frac{8-4}{8} = \frac{1}{2}$. Thus, the probability of distributing $k_1 = 4$ ones to the columns without 0's is larger than or equal to $(1 - \tilde{\alpha})^{k_1} = \frac{1}{16}$. Thus, the probability that a random function can be represented by using only x_1, x_2 and x_3 is $\eta \geq \frac{1}{16}$. Note that for this particular function, the column for $(x_1, x_2, x_3) = (1, 0, 0)$ has both 0 and 1. Thus, this function depends on $\{x_4, x_5, x_6\}$. ■

Lemma 5.1.2 considers the probability for one partition: $X_1 = (x_1, x_2, \ldots, x_p)$ and $X_2 = (x_{p+1}, x_{p+2}, \ldots, x_n)$. However, in practice, we can use the best partition of variables to represent the function. The following lemma considers such a case:

Lemma 5.1.3 *Consider a set of random functions, where k_0 combinations are mapped to 0, k_1 combinations are mapped to 1, and the other $2^n - (k_0 + k_1)$ combinations are undefined. Then, the probability that f can be represented by using only p variables is at least*

$$1 - \sigma^{\binom{n}{p}},$$

where $\sigma = 1 - \eta$, and η is the probability that a function can be represented by using only $x_1, x_2, \ldots, x_{p-1}$ and x_p.

(Proof) The probability that a function cannot be represented by using $x_1, x_2, \ldots, x_{p-1}$ and x_p is $\sigma = 1 - \eta$. Since there are $\binom{n}{p}$ ways to choose p variables out of n variables, the probability that a function cannot be represented by using any combinations of p variables is $\sigma^{\binom{n}{p}}$, by the IID assumption. The probability that a function can be represented by using at least one combination of p variables is $1 - \sigma^{\binom{n}{p}}$.   □

Example 5.1.2 In the decomposition chart in Fig. 5.1, the probability that a function cannot be represented by x_1, x_2 and x_3 is $\sigma = 1 - \eta = \frac{15}{16}$. Since there are $\binom{6}{3} = 20$ ways to choose 3 variables out of 6 variables, the probability that a function cannot be represented by any combination of 3 variables is $(\frac{15}{16})^{20} \simeq 0.275$, if we use the IID assumption.

The probability that a function can be represented by using at least one combination of 3 variables is at least $1 - \sigma^{\binom{6}{3}} = 0.725$. By the way, for this function, 3 variables are sufficient to represent the function: (x_2, x_3, x_6).   ■

From Lemma 5.1.3, we have the following:

Theorem 5.1.1 *Consider a set of random functions, where k_0 combinations are mapped to 0, k_1 combinations are mapped to 1, and the other $2^n - (k_0 + k_1)$ combinations are undefined. If*

$$p \geq \lceil \log_2(k_0 k_1) \rceil - 2,$$

then more than 95% of the functions can be represented with p variables.

(Proof) Since $\sigma = 1 - \eta < 1.0$, $1 - \sigma^{\binom{n}{p}}$ approaches 1.0, as n increases. When $p < n$, $\binom{n}{p} \geq n(n-1)/2$. Assume that $n \geq 20$. The condition that $\sigma^{\binom{n}{p}} \leq 0.05$ implies $\sigma < 0.984$. Thus, if $\eta \geq 0.0156$, then at least 95% of the functions can be realized with p variables. When $\tilde{\alpha}_0$ is sufficiently small, $1 - \tilde{\alpha}_0$ is approximated by $e^{-\tilde{\alpha}_0}$. Thus,

$$\eta \geq (1 - \tilde{\alpha}_0)^{k_1} \simeq e^{-\tilde{\alpha}_0 k_1} = e^{-\frac{k_0 \cdot k_1}{2^p}}.$$

When $p \geq \log_2(k_0 k_1) - 2$, we have $\eta > e^{-4} = 0.0183$.   □

There exist partially defined functions that require all the variables, but the fraction of such functions approaches to zero.

Theorem 5.1.2 *Consider a set of random functions, where k_0 combinations are mapped to 0, k_1 combinations are mapped to 1, and the other $2^n - (k_0 + k_1)$ combinations are undefined. If $k_0 + k_1 \ll 2^n$, then the fraction of the functions represented with $p = \lceil \log_2(k_0 k_1) \rceil - 2$ variables approaches 1.0, as n increases.*

Lemma 5.1.4 *Consider a set of random functions, where k_0 combinations are mapped to 0, k_1 combinations are mapped to 1, and the other $2^n - (k_0 + k_1)$ combinations are*

undefined. The probability that the function f requires more than p variables is $\sigma^{\binom{n}{p}}$, where
$p = \lceil \log_2(k_0 k_1) \rceil - 2$.

Theorem 5.1.3 *Consider a set of random functions, where k_0 combinations are mapped to 0, k_1 combinations are mapped to 1, and the other $2^n - (k_0 + k_1)$ combinations are undefined. When n is large, the fraction of the functions that require more than p variables to represent the function approaches zero, where $p = \lceil \log_2(k_0 k_1) \rceil - 2$.*

(Proof) It is easy to verify that $\sigma^{\binom{n}{p}} \to 0$, as $n \to \infty$. □

Lemma 5.1.5 *Suppose that a box contains 2^n balls and among them, u balls are black showing true minterms. Assume that a truth table of a random function of n variables is generated by Algorithm 5.1.1. If both n and u are sufficiently large, and $u \ll 2^n$, then for most functions generated by Algorithm 5.1.1, the numbers of true minterms are near u.*

(Explanation) By the **law of large numbers**, the numbers of true minterms in the generated truth tables form the **normal distribution**, as shown in Fig. 5.2. Note that the expected number of true minterms is $\mu = u$, and the **standard deviation** is approximately $\sigma = \sqrt{u}$. The probability that the generated truth table has u_0 true minterms, where $u - 2\sigma$ and $u + 2\sigma$ is 0.9545 [4]. □

Let $n = 1000$ and $u = 2^{20}$, then $2\sigma = 2^{11}$. In the case of 95.45% of the functions, the numbers of true minterms are between $u - 2^{11}$ and $u + 2^{11}$.

Fig. 5.2 Normal distribution

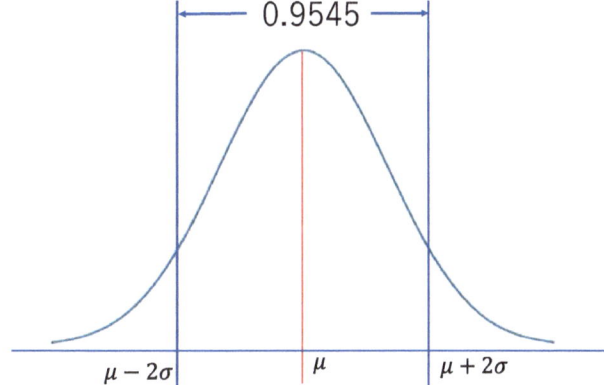

5.2   Experimental Results

In this part, we consider SOPs for various two-class functions. Here, we use ESPRESSO-MV [5] as the SOP minimizer. With the *.type fr* option, the program reads the ON set and the OFF set, and produces an SOP with reduced products for the ON set. In this case, the complement of $ON \cup OFF$ is computed to generate the DC set, which is used to expand the products.

5.2.1   Random 6-Variable Functions with $k_0 = k_1 = 4$

We randomly generated 10000 functions of 6 variables, with $k_0 = k_1 = 4$, and minimized the numbers of variables by Algorithm 3.2.1.

Table 5.1 shows the distribution of the number of variables. To represent 6-variable functions with $k_0 = k_1 = 4$, on the average, 2.89 variables are necessary. This table shows that the fraction of the functions that can be represented with at most 3 variables is 0.8325. Note that Example 5.1.2 shows that this value is at least 0.725.

5.2.2   Number of Functions That Require n_o Variables, for $n = 16, 20, 24,$ 28 and $k_0 = k_1 = 64$

We randomly generated 10,000 functions of 16, 20, 24, and 28 variables, with $k_0 = k_1 = 64$, and minimized the numbers of variables by Algorithm 3.2.1. Table 5.2 shows the number of functions that require n_o variables to represent these functions. When $n \geq 20$, most functions can be represented with 9 or 8 variables.

Note that $\lceil \log_2(k_0 k_1) \rceil - 2 = 12 - 2 = 10$. Thus, the properties of Theorems 5.1.1 and 5.1.2 are satisfied. Also, with the increase of n, the average number of variables decrease. Compared with the case of $n = 6$ and $k_0 = k_1 = 64$, the variance is very small when $n \geq 20$.

5.2.3   Random 24-Variable Functions with $k_0 = k_1$

For different values of $k_0 = k_1 = u$, where $k_1 = |ON|$ and $k_0 = |OFF|$, we generated random functions of 24 variables, and by Algorithm 3.2.1, we obtained all possible irredundant

Table 5.1 Number of functions that require n_o variables when $n = 6$ and $k_0 = k_1 = 4$

	$n_o = 1$	$n_o = 2$	$n_o = 3$	$n_o = 4$	$n_o = 5$
# Functions	515	1816	5994	1606	69

Table 5.2 Number of functions that require n_o variables. for $n = 16, 20, 24, 28$ and $k_0 = k_1 = 64$. (Each row corresponds to 10,000 functions)

n	$n_o = 8$	$n_o = 9$	$n_o = 10$	$n_o = 11$	Average
16	17	7828	2152	3	9.2141
20	197	9787	16	0	8.9819
24	574	9426	0	0	8.9426
28	1001	8999	0	0	8.8999

Table 5.3 Distribution of irredundant solutions requiring n_o variables. (Each column corresponds to single function)

n_o	$k_0 = k_1 = u$						
	16	32	64	128	256	512	1024
5	57						
6	2240						
7	15913	70					
8	7576	10455					
9	231	51788	116				
10		8469	15310				
11		19	62513	1219			
12			6872	48286			
13			30	47917	1287		
14				902	35638		
15					22976	154	
16					147	5826	
17						5127	125
18						31	2027
19							523
20							2

sets of variables. Table 5.3 shows the distribution of irredundant solutions that require n_o variables for different values of u. With the increase of u, the number of variables tends to increase. In this table, we can observe that the numbers of variables are between $2 \log_2 u - 3$ and $2 \log_2 u + 1$. Note that these results are consistent with Theorem 5.1.1.

Table 5.3 shows that when $k_0 = k_1 = 64$, to represent a 24-variable function, 9 variables are necessary. Thus, when an exact algorithm, such as Algorithm 3.2.1, is used to minimize the variables for this function, we have a solution with $n_o = 9$ variables. On the other hand, when a heuristic algorithm, such as Algorithm 4.2.1, is used, in the worst case, we have a

Table 5.4 Number of variables to represent 24-variable random functions with different values of (k_0, k_1). (Each entry shows average of 100 functions)

k_0	k_1									
	10	20	30	40	50	60	70	80	90	100
10	**3.96**	4.90	5.08	5.77	5.99	6.02	6.25	6.62	6.86	6.96
20	4.91	**5.83**	6.04	6.57	6.93	7.00	7.07	7.25	7.68	7.89
30	5.05	6.04	**6.85**	6.99	7.20	7.73	7.98	8.00	8.00	8.04
40	5.82	6.52	7.00	**7.31**	7.92	8.00	8.02	8.23	8.61	8.85
50	5.91	6.96	7.23	7.97	**8.00**	8.06	8.41	8.88	8.97	9.00
60	5.99	7.00	7.70	7.98	8.07	**8.65**	8.94	8.98	9.00	9.04
70	6.16	7.05	7.98	8.00	8.53	8.94	**8.98**	9.00	9.03	9.40
80	6.62	7.37	7.99	8.22	8.91	9.00	9.00	**9.11**	9.50	9.90
90	6.85	7.83	8.00	8.69	8.99	9.00	9.05	9.53	**9.82**	9.99
100	6.92	7.93	8.06	8.84	9.00	9.01	9.42	9.85	9.99	**10.00**

solution with $n_o = 13$ variables, while in more than 90% of cases, we have a solution with at most $n_o = 11$ variables.

By the way, Table 5.2 shows that when $n = 24$, 9426 functions out of 10,000 functions require 9 variables.

5.2.4   Random 24-Variable Functions with $k_0 \neq k_1$

For each pair of (k_0, k_1), we generated 100 random functions of 24 variables, applied Algorithm 3.2.1, and obtained numbers of variable to represent functions. Table 5.4 shows the average numbers of variable to represent the functions for different pairs (k_0, k_1). From this table, we can observe that with the increase of k_0 and k_1, the numbers of variables increase. Also, we can observe that the table is symmetric with respect to the main diagonal elements, which are denoted by bold numbers. Figure 5.3 plots the relation between $k_0 k_1$ and the number of variables. The orange triangles show the upper bound obtained by Theorem 5.1.1: $\log_2(k_0 k_1) - 2$, while the blues plots show the experimental results. Note that the horizontal axis is $\log_2$ scale. We can confirm that the results are consistent with Theorem 5.1.1.

5.2.5   Random 60-Variable Function

We generated a random function with $n = 60$ and $k_0 = k_1 = u = 8192$. In this case, derivation of all possible minimal sets of variables was too time-consuming.

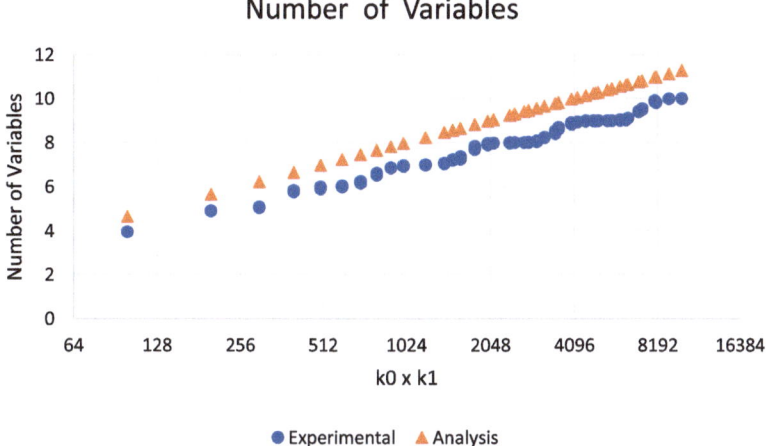

Fig. 5.3 Number of variables versus $k_0 k_1$ for 24-variable functions

ESPRESSO-MV produced a 757-product solution with 60 variables, that is, the SOP requires all the variables. When the number of variables is first reduced by Algorithm 4.2.1, we obtained a $n_o = 24$-variable solution, and then by ESPRESSO-MV, we obtained an SOP with 1105 products. Recall, the PLA measure is $n \cdot t$. Thus,

- By SOP minimization only
$$60 \times 757 = 45,420.$$
- By variable minimization and then by SOP minimization

$$24 \times 1,105 = 26,520.$$

Thus, for this function, by first reducing variables, and then by reducing products, we have a PLA with a lower PLA measure.

5.2.6 Non-random Functions

As for non-random functions, we used *Connect-4* from the UCI (University of California-Irvine) data set [6].Connect-4 is a two-player board game. Two players, X and Y, drop disks into a vertically suspended 6×7 grid. The number of squares is $n = 6 \times 7 = 42$. Each variable takes $q = 3$ values: empty, X, and Y. The number of classes is three: (1) player X wins, (2) player Y wins, and (3) draw. The number of instances (i.e., care minterms) is 67,557. Among these, player X wins in 44,473 instances, player Y wins in 16,635 instances,

Table 5.5 Numbers of variables and products for Connect4

Function	Optimization	Variables	Products	PLA measure
1	SOP	121	1,906	230,626
	Var+SOP	61	2,371	144,631
2	SOP	123	1,610	198,030
	Var+SOP	62	1,925	119,350
3	SOP	124	1,608	199,392
	Var+SOP	61	1,909	116,449

and a draw occurs in 6,449 instances. Assume that **one-hot encoding**[3] is used to represent a 3-valued variable, the number of bits to represent the input part is $n = 42 \times 3 = 126$.

From this, we generated three functions:

1. Player X wins or not,
2. Player Y wins or not, and
3. The game is draw or not.

Table 5.5 shows the results. The upper lines show the result, when only SOP minimizations (ESPRESSO-MV) were applied. For the first function, the number of products was reduced to 1,906. At the same time, the number of variables was reduced to 121. The lower lines show the results, when Algorithm 4.2.1 was applied first, and the SOP minimization was applied second. For the first function, the number of variables was reduced to 61, and the number of products was reduced to 2,371. The last column shows the PLA measure. This shows that the PLA measure can be reduced by first minimizing the variables. For other functions, we had similar results.

Next, check if Theorem 5.1.1 is applicable or not to the first function. Since $k_1 = 44473$ and $k_0 = 16635 + 6449$, we have $UB = \lceil \log_2(k_0 k_1) \rceil - 2 = 58$. Thus, Theorem 5.1.1 is not applicable to this function. Note that this function is not random.

Also for sparse multi-valued input multi-class functions, there is a trade-off between the number of variables and the number of products [7].

5.2.7 MNIST Handwritten Character Recognition Circuit

MNIST [8] is a data set of handwritten digits. The training set consist of 6×10^4 images. Selected images are shown in Fig. 5.4. Each image is a bit map of 28×28 pixels. Originally, each pixel is represented by an 8-bit number (i.e., a grayscale image of 256 values), but we

---

[3] In this encoding, p bits are used to represent a p-valued variable. For example, in a three-valued variable, 100 denotes 1; 010 denotes 2; and 001 denotes 3.

Fig. 5.4 MNIST handwritten character recognition [9, 10]

converted it into binary number by setting the threshold 96. There are 10 different digits. Thus, we had an n-variable m-valued classification function, where $n = 28 \times 28 = 784$, and $m = 10$. Also in this process, we removed duplicated data.

Figure 5.5 shows the **45-unit realization** [11], where for each pair of digits, we use a **unit**. In the circuit, each unit decides if the input image represents the digit i, or the digit j, or another digit. With $\binom{10}{2} = 45$ such units, we can make a final decision using counters. The counter with the largest value identifies the most probable input digit.

Fig. 5.5 45-unit realization for MNIST. © 2021 IEICE. Reprinted with permission from [11]

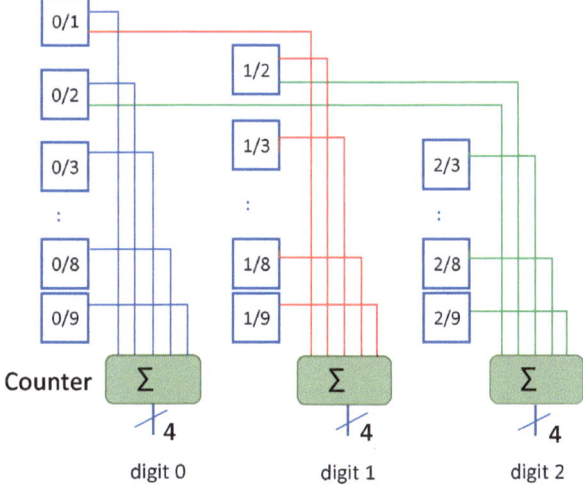

Table 5.6 Number of variables to distinguish a pair of digits (i,j) in MNIST

	1	2	3	4	5	6	7	8	9
0	13	19	17	16	20	20	17	17	17
1		21	18	18	17	17	21	20	18
2			21	20	19	20	23	21	20
3				18	23	18	20	22	21
4					19	19	20	19	23
5						21	20	21	20
6							16	20	18
7								19	**25**
8									22

In Fig. 5.5, a square symbol denotes a unit. The unit i/j has two outputs: The output $(1, 0)$ denotes that the input image represents the digit i; the output $(0, 1)$ denotes that the input image represents the digit j; and the output $(0, 0)$ denotes that the input image represents another digit or unknown. Thus, each unit produces a **ternary output**. Since there are 45 units, the total number of outputs is 90. Figure 5.5 shows the circuits for only digits 0, 1 and 2. Circuits for other digits are omitted. In addition, we use 10 **counters**, shown by $\sum$ symbols in Fig. 5.5. A counter counts the number of 1's in the inputs. It can be combinational or sequential. The ith counter has 9 inputs with label i, and counts the number of 1's in the inputs, and represents it by a 4-bit binary number. To estimate the size of each unit, we use Theorem 5.1.2. Since the number of images for each digit is approximately 6000, we have

$$UB = \lceil \log_2(6000 \times 6000) \rceil - 2 = \lceil 23.1 \rceil = 24.$$

Thus, we can estimate that most units can be realized with 24 variables.

We reduced the numbers of variables to represent each unit by Algorithm 4.2.1. Table 5.6 shows the numbers of variables to represent the units. Since Algorithm 4.2.1 is a heuristic one, the number of variables may not be exactly minimum.

Among 45 units, except for the unit to distinguish digits 7 from 9, each unit can be represented with at most 23 variables. To distinguish digits 7 from 9, we need 25 variables. However, with a linear transformation[4] with compound degree two, digits 7 and 9 can be distinguished using only 18 variables.

To distinguish digits 0 from 1 is easy: 13 variables are sufficient.

---

[4] Details are shown in Chap. 6.

5.3   Remarks

This chapter considered the reduction of variables for partially defined two-class functions, where the ON set and the OFF set are much smaller than the DC set (i.e., sparse functions). Also, it showed that there is a trade-off between the number of variables and the number of products in the SOPs. When the given function is sparse, the minimization of the variable first, and then the minimization of the products often reduces the PLA measure.

For randomly generated two-class functions with k_0 false minterms and k_1 true minterms, the number of variables can be reduced to $\lceil \log_2(k_0 k_1) \rceil - 2$. We derived this result by using the IID assumption.

This chapter is based on [11, 12].

5.4   Exercises

5.1 (E) Consider the 45-unit realization for MNIST in Fig. 5.5. Show that the training accuracy is 1.00.

5.2 (E) Consider the randomly generated classification function:

$$f : \{0, 1\}^{60} \rightarrow \{0, 1, 2\}.$$

Let the number of vectors $\vec{a}$ such that $f(\vec{a}) = i$ be 1000 for $i = 0, 1, 2$.

1. Design the 3-unit realization of the classifier.
2. Estimate the number of variables for each unit.

References

1. Ashenhurst RL (1957) The decomposition of switching functions. In: International symposium on the theory of switching, pp 74–116
2. https://en.wikipedia.org/wiki/Independent_and_identically_distributed_random_variables
3. Bishop CM (2006) Pattern recognition and machine learning. Springer
4. https://en.wikipedia.org/wiki/Standard_deviation
5. Rudell RL, Sangiovanni-Vincentelli A (1987) Multiple-valued minimization for PLA optimization. IEEE Trans CAD 6(5):727–750
6. https://archive.ics.uci.edu/ml/datasets/Connect-4
7. Sasao T (2022) A method to generate classification rules from examples. ISMVL, May 18–22, online, pp 176–181
8. http://yann.lecun.com/exdb/mnist/
9. https://commons.wikimedia.org/wiki/File:MnistExamples.png
10. CC BY-SA 4.0 (2017) MnisExample.png, Created:14 Dec 2017

11. Sasao T, Horikawa Y, Iguchi Y (2021) Classification functions for handwritten digit recognition. IEICE Trans Inf Syst E104-D(8):1076–1082
12. Sasao T (2022) Two-level minimization for partially defined functions. IWLS, online, July 18–20

Linear Decomposition

6

This chapter introduces linear transformations of the input variables. With this technique, the number of variables can be reduced drastically. Experimental results using MNIST[1] and fashion MNIST[2] data sets are also shown.

6.1 Algorithm for Linear Decomposition

The number of variables in a partially defined function often can be reduced by a **linear decomposition** [1], shown in Fig. 6.1, where L realizes a **linear function**, while G realizes a **non-linear function**. The cost of the **linear part** is $O(np)$, while the cost of the **non-linear part** is $O(q2^p)$.

Definition 6.1.1 A **compound variable** has the form $y = c_1x_1 \oplus c_2x_2 \oplus \cdots \oplus c_nx_n$, where $c_i \in \{0, 1\}$. The **compound degree** of a variable y is $\sum_{i=1}^{n} c_i$, where $\sum$ denotes an ordinary integer addition, and c_i is treated as an integer. A **primitive variable** is a variable whose compound degree is one.

Definition 6.1.2 Given a partially defined function f, a **linear transformation** that minimizes p, the number of compound variables, is an **optimal transformation**.

When the number of the compound variables is reduced to $q = \lceil \log_2 m \rceil$, the linear transformation is optimum by Theorem 3.1.2.

---

[1] MNIST is the set of images for handwritten digits. See Sect. 5.2.7.
[2] Fashion MNIST is the set of images for fashion items. See Sect. 6.4.5.

T. Sasao, *Classification Functions for Machine Learning and Data Mining*, Synthesis Lectures on Digital Circuits & Systems, https://doi.org/10.1007/978-3-031-35347-5_6

Fig. 6.1 Linear decomposition

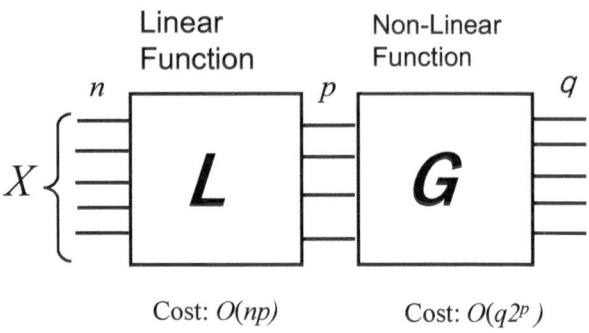

Cost: $O(np)$ Cost: $O(q2^p)$

When a partially defined function satisfies a certain condition, there exists a linear transformation that reduces the number of variables.

Lemma 6.1.1 *[2] An n-variable partially defined function*

$$f(x_1, x_2, \ldots, x_n) = (F_1, F_2, \ldots, F_m)$$

can be represented by using at most $n - 1$ compound variables, if and only if there exists a non-zero vector $\vec{u}$ such that

$$\vec{u} \in B^n \setminus D_f,$$

where D_f is the set of difference vectors of f.

(Proof $\Leftarrow$) Without loss of generality, we can assume that there exists a vector $\vec{u}$ such that $x_1 = 1$. Let $S(F_i, x_1 = a)$, $(i = 1, 2, \ldots, m)$ be the sets of vectors in F_i such that $x_1 = a$, where $a \in \{0, 1\}$. Next, consider the following $2m$ sets of vectors:

$$A_i = S(F_i, x_1 = 0)$$
$$B_i' = \{\vec{b} \oplus \vec{u} \mid \vec{b} \in S(F_i, x_1 = 1)\}$$

In this case, f can be represented with $n - 1$ compound variables. The first bits of A_i, $(i = 1, \ldots, m)$ are all 0's. Also, the first bits of B_i', $(i = 1, \ldots, m)$ are all 0's. Thus, the first bit of each vector is 0 in these $2m$ sets of vectors.

From the property of registered vectors, all the vectors in A_i, $(i = 1, \ldots, m)$ are distinct. Similarly, all the vectors in B_i', $(i = 1, \ldots, m)$ are distinct.

Next, we will show that all the vectors in A_i and B_j', $(i \neq j)$ are distinct.

On the contrary, assume that a vector $\vec{a}$ in A_i, and a vector $\vec{b}'$ in B_j' are equal. Then, we have the relation $\vec{a} = \vec{b}'$. Since $\vec{b}' = \vec{b} \oplus \vec{u}$, we have a relation $\vec{a} = \vec{b} \oplus \vec{u}$, and the relation $\vec{u} = \vec{a} \oplus \vec{b}$. However, this contradicts the condition of $\vec{u}$. Thus, any vector in A_i and any vector in B_j', $(i \neq j)$ are different. To summarize, the first bits of

Table 6.1 Registered vector table

	x_1	x_2	x_3	x_4	x_5	x_6	f
$\vec{a}_1$	1	1	0	0	1	1	1
$\vec{a}_2$	0	1	1	0	1	1	1
$\vec{a}_3$	0	0	0	0	1	0	1
$\vec{b}_1$	1	1	0	1	1	1	2
$\vec{b}_2$	1	0	0	0	1	1	2
$\vec{b}_3$	0	1	0	1	0	0	2

the vectors in $A_1, A_2, \ldots, A_m, B'_1, B'_2, \ldots B'_m$ are all 0's, and all the vectors in the sets $A_1 \cup B'_1, A_2 \cup B'_2, \ldots, A_m \cup B'_m$ are distinct.

Adding the vector $\vec{u}$ to an element in the set $S(F_i, x_1 = 1)$ corresponds to performing a linear transformation to $S(F_i, x_1 = 1)$. Thus, to represent f, the first bit x_1 is not necessary, and the remaining $n - 1$ compound variables are sufficient to represent the function.

(Proof $\Rightarrow$) We prove by contraposition. Since $B^n = D_f \cup \vec{0}$, for any non-zero $\vec{u}$ that we chose in the operation of the proof $\Leftarrow$ of Lemma 6.1.1, we have that $\vec{u} = \vec{a} \oplus \vec{b}$, where $\vec{a} \in F_i$ and $\vec{b} \in F_j$, and $i \neq j$. Since $\vec{u}$ is a non-zero vector, without loss of generality, we can assume that $x_1 = 1$ in $\vec{u}$. Hence, in $\vec{a}$, $x_1 = 0$ and in $\vec{b}$, $x_1 = 1$ (or vice-versa), which means that $\vec{a} \in A_i$ and $\vec{a} = \vec{b}' = \vec{u} \oplus \vec{b} \in B'_j$, implying that $\vec{u}$ cannot be used to omit variable x_1. Thus, we cannot omit any variable.   □

Theorem 6.1.1 *A necessary and sufficient condition that an n-variable function* $(F_1, F_2, \ldots, F_m)$ *is represented using at most* $n - 1$ *compound variables is the number of vectors in the set of difference vectors of f is less than* $2^n - 1$.

(Proof) This is a direct consequence of Lemma 6.1.1.   □

Algorithm 6.1.1 (Reduction of Compound Variables)

1. Derive the set of difference vectors D_f of an n-variable function.
2. If $|D_f| = 2^n - 1$, then stop, since reduction is impossible.
3. Obtain a non-zero vector $\vec{d} \in B^n \setminus D_f$ with the minimum weight.[3]
4. Remove one variable from $\vec{d}$, and apply the linear transformation to the function.
5. Let $n \leftarrow n - 1$, and go to step 1.

Example 6.1.1 Let us reduce the number of variables in the classification function shown in Table 6.1, which is the same function as the one in Table 3.1.

---

[3] The **weight of a vector** is the number of 1's in the vector.

Table 6.2 Difference vectors generated from Table 6.1

x_1	x_2	x_3	x_4	x_5	x_6	TAG
0	0	0	1	0	0	$(\vec{a}_1, \vec{b}_1)$
0	1	0	0	0	0	$(\vec{a}_1, \vec{b}_2)$
1	0	0	1	1	1	$(\vec{a}_1, \vec{b}_3)$
1	0	1	1	0	0	$(\vec{a}_2, \vec{b}_1)$
1	1	1	0	0	0	$(\vec{a}_2, \vec{b}_2)$
0	0	1	1	1	1	$(\vec{a}_2, \vec{b}_3)$
1	1	0	1	0	1	$(\vec{a}_3, \vec{b}_1)$
1	0	0	0	0	1	$(\vec{a}_3, \vec{b}_2)$
0	1	0	1	1	0	$(\vec{a}_3, \vec{b}_3)$

1. Table 6.2 shows the set of difference vectors. The last column shows the pair of registered vectors that produced the difference vector.
2. $|D_f| = 9 < 2^6 - 1$.
3. Select $\vec{d} = (0, 0, 0, 0, 0, 1)$ as the vector with the minimum weight.
4. Delete the variable x_6. $n \leftarrow 5$.
5. Update the set of difference vectors. (Delete the column for x_6 from Table 6.2).
6. $|D_f| = 9 < 2^5 - 1$.
7. Select $\vec{d} = (0, 0, 0, 0, 1)$ as the vector with the minimum weight.
8. Delete the variable x_5. $n \leftarrow 4$.
9. Update the set of difference vectors. (Delete the columns for x_5 and x_6 from Table 6.2).
10. $|D_f| = 9 < 2^4 - 1$.
11. Select $\vec{d} = (0, 0, 1, 0)$ as the vector with the minimum weight.
12. Delete the variable x_3. $n \leftarrow 3$.
13. Update the set of difference vectors. (Delete the columns for x_3, x_5, and x_6 from Table 6.2).
14. Stop the algorithm, since the number of vectors in the set of difference vector is $2^3 - 1 = 7$. The function can be represented with $\{x_1, x_2, x_4\}$.   ∎

Example 6.1.2 Let us reduce the number of variables of the classification function shown in Table 6.1. In this case, we show that the order of the selection of the variables influences the results.

1. In Example 6.1.1, assume that we first delete x_1, x_3, and x_5, and go to the step 12. The remaining variables are x_2, x_4 and x_6. Table 6.3 shows the registered vector table.
2. In this case, the set of difference vectors does not contain $\vec{d} = (1, 0, 1)$, we select $\vec{d} = (1, 0, 1)$.

Table 6.3 Reduced
registered vector table

x_2	x_4	x_6	g
1	0	1	1
1	0	1	1
0	0	0	1
1	1	1	2
0	0	1	2
1	1	0	2

Table 6.4 Minimized
registered vector table

$y_1 = x_4$	$y_2 = x_2 \oplus x_6$	g
0	0	1
0	0	1
0	0	1
1	0	2
0	1	2
1	1	2

3. Let x_2 be a pivot variable, and we have a linear transformation that converts the registered vector table into Table 6.4.
4. In this case, the function can be represented with two variables: $y_1 = x_4$ and $y_2 = x_2 \oplus x_6$.
5. The function can be represented as $g = \bar{y}_1 \bar{y}_2 \vee 2(y_1 \vee y_2)$. ∎

Example 6.1.3 Consider the function shown in Table 2.3. Find the linear decomposition for this function using Algorithm 6.1.1.

From Table 2.3, we have the set of difference vectors in Table 6.5.
Note that $(x_1, x_2, x_3, x_4) = (1, 1, 0, 0)$ is missing in Table 6.5. So, we perform the linear transformation: $y_1 = x_1 \oplus x_2$, and have Table 6.6, where * denote duplicated rows.
Note that $(y_1, x_3, x_4) = (0, 1, 1)$ is missing in Table 6.6. So, we perform the linear transformation $y_2 = x_3 \oplus x_4$, and have the registered vector table shown in Table 6.7, where * denotes the duplicated row.
Table 6.8 shows the difference vector table derived from Table 6.7. Note that the vector $(y_1, y_2) = (1, 1)$ is missing in Table 6.8. So, we perform the linear transformation: $z_1 = y_1 \oplus y_2 = x_1 \oplus x_2 \oplus x_3 \oplus x_4$. Thus, the function in Table 2.3 can be represented by z_1. ∎

Table 6.5 The set of difference vectors for the function in Table 2.3

x_1	x_2	x_3	x_4	TAG
0	0	0	1	(1,5)
0	1	1	0	(1,6)
0	1	0	0	(1,7), (2,8)
1	1	1	0	(1,8),(2,7),(3,6)
1	0	1	1	(2,5),(4,6)
1	1	0	1	(2,6),(3,7),(4,5)
1	0	0	0	(3,5),(4,7)
0	1	1	1	(3,8)
0	0	1	0	(4,8)

Table 6.6 The set of difference vectors after linear transformation: $y_1 = x_1 \oplus x_2$

y_1	x_3	x_4	TAG
0	0	1	(1,5)
1	1	0	(1,6)
1	0	0	(1,7)
0	1	0	(1,8)
1	1	1	(2,5)
0	0	1	(2,6)*
1	0	0	(3,5)*
1	1	1	(3,8)*
0	1	0	(4,8)*

6.2   Number of Variables to Represent a Non-linear Function

This section considers a method to estimate p, the number of variables to represent a non-linear function in Fig. 6.1.

Definition 6.2.1 A classification function f is **reducible** if f can be represented with fewer variables than the original function when a linear decomposition is used. Otherwise, f is **irreducible**.

Theorem 6.2.1 *An n-variable classification function f is reducible if and only if the set of difference vectors for f contains fewer than $2^n - 1$ elements.*

(Proof) This is a direct consequence of Lemma 6.1.1.   □

Table 6.7 Registered vector table derived from Table 2.3 by linear transformations: $y_1 = x_1 \oplus x_2$, $y_2 = x_3 \oplus x_4$

y_1	y_2	f	TAG
0	0	1	1
1	1	1	2
1	1	1	3*
0	0	1	4*
0	1	2	5
1	0	2	6
1	0	2	7*
0	1	2	8*

Table 6.8 The set of difference vectors after linear transformations: $y_1 = x_1 \oplus x_2$, $y_2 = x_3 \oplus x_4$

y_1	y_2	TAG
0	1	(1,5), (2,6)
1	0	(1,6), (2,5)

Lemma 6.2.1 Let $f = (F_1, F_2, \ldots, F_m)$ be a classification function. Let N be the number of distinct difference vectors. Then,

$$N \le \sum_{(i<j)} k_i k_j,$$

where $i, j \in \{1, 2, \ldots, m\}$, and $k_i = |F_i|$.

Theorem 6.2.2 Let N be the number of distinct difference vectors of a classification function f. Then, f can be represented with at most $p = \lfloor \log_2(N+1) \rfloor$ compound variables.

(Proof) By Theorem 6.2.1, if $N < 2^n - 1$, then we can omit one variable. With repeated application of this theorem, the number of variables can be reduced to $p = \lfloor \log_2(N+1) \rfloor$. When $N + 1 = 2^p$, the function can be represented with p variables. □

The above theorem gives an upper bound on p, the number of variables for the non-linear part. Unfortunately, it is not tight. This is because the difference vectors are modified by the linear transformations, and some of them become identical.

6.3 Examples of Reductions

Example 6.3.1 In the world, there are 197 independent countries [3]. The continent of these countries can be classified into $m = 6$ areas: Europe, Asia, Africa, North America, South America, and Oceania.

Table 6.9 Minimization results for Examples 6.3.1–6.3.4. © 2020 IEEE. Reprinted with permission from [2]

Example	n	m	k	AL 3.2.1	AL 3.2.1 + AL 6.1.1	AL 6.1.1
6.3.1	60	6	197	12	10	9
Countries				(1)	(3)	(11)
6.3.2	60	7	118	10	9	8
Elements				(1)	(2)	(12)
6.3.3	30	9	3700	20	18	18
Companies				(1)	(3)	(4)
6.3.4	30	4	4000	20	18	18
Random				(1)	(3)	(3)

The number in the parenthesis denotes the maximum compound degree

Now, consider the classification function that produces the area for each country. The names of the countries are described by alphabets (26 characters), blank and hyphen(-). The country with the longest name is "Saint Vincent and the Grenadines," having 32 characters. However, all the countries can be distinguished by the first 12 characters.

Thus, to show the countries, we used the first 12 characters. All the upper case letters are converted into lower case letters, and all the characters are coded by 5 bits. For the countries whose name have less than 12 characters, blank characters were appended to make the lengths of all the names 12.

In the original classification function, the number of inputs is $n = 5 \times 12 = 60$, the number of outputs is $m = 6$ (one-hot encoding), and the number of registered vectors is $k = 197$. Among these, Armenia, Azerbaijan, Cyprus, Georgia, Kazakhstan, Turkey, Russia, are classified as part of Europe.

1. Minimizing the primitive variables by Algorithm 3.2.1, yields a solution with 12 variables.
2. Minimizing the compound variables by Algorithm 6.1.1 to the result of 1), yields a solution with 10 variables. In this case, the maximum compound degree was three.
3. Minimizing the compound variables by Algorithm 6.1.1 directly, yields a solution with 9 variables. In this case, the maximum compound degree was 11. ∎

Table 6.9 summarizes the experimental results. The column headed with AL 3.2.1 denotes the number of the primitive variables obtained by Algorithm 8.2.1. The column headed with AL 3.2.1 + AL 6.1.1 denotes the number of the compound variables obtained by Algorithm 3.2.1 and by the iterative algorithm using Lemma 6.1.1. The column headed with AL 6.1.1 denotes the number of the compound variables obtained by Algorithm 6.1.1. The number in the parenthesis shows the maximum compound degree.

Example 6.3.2 Up to now, 118 chemical elements are known [4]. These chemical elements can be classified into $m = 7$ periods according to their number of protons. For example, Hydrogen is in the first period; Lithium is in the second period; Sodium is in the third period; Potassium is in the fourth period; Rubidium is in the fifth period; Cesium is in the sixth period; and Francium is in the seventh period.

Consider the classification function that gives the corresponding period for a given name of a chemical element. The name of the chemical elements is represented by 13 alphabet characters. However, all the elements can be distinguished by the first 12 characters. Thus, in the classification functions, we consider only the first 12 characters. Similarly to Example 6.3.1, each character is coded by 5 bits. In the original classification function, the number of inputs is $n = 5 \times 12 = 60$, the number of outputs is $m = 7$ (one-hot encoding), and the number of registered vectors is $k = 118$. We minimized the variables in the same way as in Example 6.3.1. The second row of Table 6.9 shows the experimental results.   ∎

Example 6.3.3 Consider the classification function, where the inputs are the telephone numbers of 3700 Japanese companies [5], and the outputs are the names of the stock exchange. There are $m = 9$ different stock exchange. (1) Tokyo 1st; (2) Tokyo 2nd; (3) Tokyo Mothers; (4) Sapporo; (5) Nagoya; (6) Fukuoka; (7) JASDAQ; (8) REIT; and (9) Foreign.

The telephone numbers are represented by 9-digit decimal numbers. We converted them to binary numbers of 30-bits. Thus, the number of inputs is $n = 30$. Also, the number of outputs is 9 (one-hot encoding), and the number of registered vectors is $k = 3700$. We minimized the variables in the same way as in Example 6.3.1. The third row of Table 6.9 shows the experimental results.   ∎

Example 6.3.4 We generated 4000 distinct random vectors of 30 bits, and partitioned them into $m = 4$ sets, each consists of 1000 vectors. From this, we made a classification function with $n = 30$ inputs, four outputs, and 4000 registered vectors. We minimized the variables in the same way as in Example 6.3.1. The last row of Table 6.9 shows the experimental results.   ∎

Example 6.3.4 can be used as a packet filter of the internet.

Table 6.10 Benchmark functions

Data	# of inputs	# of classes	# of vectors
	n	m	k
$MNIST(8 \times 8)$	64	10	3,686
$MNIST(14 \times 14)$	196	10	58,191
$MNIST(28 \times 28)$	784	10	59,981
$CIFAR(32 \times 32)$	1024	2	9,930

6.4   Experimental Results for Larger Problems

6.4.1   Benchmark Functions and Computer Environment

We used the handwritten digits data sets (MNIST) represented by 8×8[4] 14×14, and 28×28 bit images [7, 8], and 32×32 images of CIFAR-10 [9]. The first three represent 10 classes. Since CIFAR-10 consists of colored images, only the blue component of the training set was binarized using the most significant bits. Also, only the first two classes are used: airplane and automobile.[5] Table 6.10 shows the number of inputs n, the number of classes m, and the number of vectors k. We used a computer with an INTEL Core i7, 7700, 3.65GHz CPU, and 64 GB main memory, on Windows 10.

6.4.2   Reduction of Primitive Variables

Table 6.11 shows the results of minimization. Algorithm 3.2.1 successfully minimized variables for 8×8 and 14×14 functions. However, for the 28×28 and 32×32 functions, it failed due to memory overflow. For the 14×14 function, the minimum covering was aborted in 10^4 seconds (i.e., 2 hours 47 minutes). The detection of essential variables was quite effective to reduce necessary memory as well as CPU time. Without detecting the essential variables, it was impossible to complete Algorithm 3.2.1 due to the memory overflow for the 14×14 function.

---

[4] MNIST(8×8) was generated from *optdigits.tra* in UCI data set [6]. Originally, it was 17-valued input, but the most significant bit was used to generate binary-input functions.

[5] We tried to reduce the variables for all 10 classes for CIFAR-10 benchmark function, but the number of compound variables exceeded 64, so we aborted the computation. Current implementation of Algorithm 6.1.1 works up to 64 compound variables. However, when only two classes are considered, the numbers of compound variables can be reduced to less than 64. Thus, we minimized the number of variables for a *two-class function*.

Table 6.11 Reduction of primitive variables

	Algorithm 3.2.1			Algorithm 4.2.1	
	Reduced	ESS	CPU	Reduced	CPU
Data	p		(s)	p	(ms)
$MNIST(8 \times 8)$	21	4	2.1	23	218
$MNIST(14 \times 14)$	41	7	9,920.4	45	28,453
$MNIST(28 \times 28)$	–	–	–	37	109,984
$CIFAR(32 \times 32)$	–	–	–	58	34,468

Table 6.12 Reduction of compound variables

	Algorithm 6.1.1		
	Reduced	CPU	Reduction of
Data	Variable: p	Time (s)	primitive variables
$MNIST(8 \times 8)$	17	2.6	Algorithm 3.2.1
$MNIST(14 \times 14)$	25	4,542.4	Algorithm4.2.1
$MNIST(28 \times 28)$	25	3,088.0	Algorithm 4.2.1
$CIFAR(32 \times 32)$	19	136.1	Algorithm 4.2.1

Algorithm 4.2.1 successfully reduced variables for four functions in a reasonable time. Algorithm 3.2.1 produced better solutions, but required much longer computation time. The column headed by ESS denotes the numbers of essential variables.

6.4.3   Reduction of Compound Variables

As for the 8×8 function, we first applied Algorithm 3.2.1 to minimize primitive variables, and then applied Algorithm 6.1.1 to obtain a 17-variable solution.

As for 16×16, 28×28, and 32×32 functions, we first applied Algorithm 4.2.1 to reduce primitive variables, and then applied Algorithm 6.1.1. Table 6.12 shows the number of compound variables and their CPU time.

6.4.4   MNIST Character Recognition Circuit (45-Unit Realization)

With a linear transformation, MNIST character recognition circuit that appeared in Sect. 5.2.7 can be simplified.

Table 6.13 Number of variables to distinguish a pair of digits in MNIST

	1	2	3	4	5	6	7	8	9
0	13	19	17	16	20	20	17	17	17
0	10	14	14	13	14	15	12	14	14
1		21	18	18	17	17	21	20	18
1		15	15	13	14	13	15	16	13
2			21	20	19	20	23	21	20
2			17	15	16	16	17	16	15
3				18	23	18	20	22	21
3				14	17	14	16	17	16
4					19	19	20	19	23
4					15	14	15	16	18
5						21	20	21	20
5						17	15	17	16
6							16	20	18
6							12	16	13
7								19	25
7								15	17
8									22
8									17

The upper figure shows the number of primitive variables
The lower figure shows the number of compound variables

Table 6.13 shows the effect of linear transformations. The upper rows show the number of the primitive variables to distinguish two digits. The lower rows show the number of compound variables to distinguish two digits, when the degrees of the compound variables were at most two.

6.4.5 Fashion MNIST (45-Unit Realization)

Figure 6.2 shows sample images of Zalando's [10] fashion items. Each image is a bit map of $28 \times 28 = 784$ pixels. Originally each pixel was represented by an 8-bit number (i.e., 256-valued grayscale). To reduce the size of data, each pixel was represented by a binary bit, where threshold=32. Similar to the case of MNIST, the data set consists of 60,000 training images, about 6000 images for each article. There are 10 items. Thus, we have a 784-input 10-output function. Thus, this function can be implemented by the circuit in Fig. 5.5, the same architecture as MNIST. Since there are 10 items, the number of pairs to distinguish items is

T-shirt/top : 0
Trouser : 1
Pullover : 2
Dress : 3
Coat : 4
Sandal : 5
Shirt : 6
Sneaker : 7
Bag : 8
Ankle boot : 9

Fig. 6.2 Fashion-MNIST data set. Reproduced from an article in ITmedia/@IT with permission [11]. Original data comes from [12, 13]

$$\binom{10}{2} = 45.$$

Table 6.14 shows the numbers of variables to distinguish different articles. The upper rows show the number of primitive variables to distinguish a pair of articles. The lower rows show the number of compound variables to distinguish a pair of articles.

Similar to the case of MNIST, since $k_i \simeq 6000$, the number of variables to distinguish a pair of different articles is $\lceil \log_2(k_i k_j) \rceil - 2 = 24$, if the images are completely random. However, to distinguish the article 2 (Pullover) and the article 4 (Coat), 55 variables are necessary when only primitive variables are used. When a linear transformation is used, the number of the compound variables is reduced to 19 variables. Note that any pair of articles can be distinguished with at most 19 compound variables. This is consistent with the upper bound

$$UP = \lfloor \log_2(k_i k_j + 1) \rfloor.$$

given by the Theorem 6.2.2.

6.4.6   Single-Unit Realization of MNIST

A **single-unit realization** is implemented by a cascade of a linear circuit and a memory, as shown in Fig. 6.3. When only primitive variables are used, the linear part can be omitted, and the function can be implemented by a single memory.

The number of primitive variables was reduced to $p_1 = 37$ by Algorithm 4.2.1. Then, the number of compound variables was reduced to $p_2 = 25$ by Algorithm 6.1.1

Table 6.14 Number of variables to distinguish a pair of articles in Fashion-MNIST

	1	2	3	4	5	6	7	8	9
0	27	33	38	30	13	50	10	28	11
0	16	18	18	17	13	19	9	16	10
1		20	32	21	9	24	5	19	9
1		14	17	14	9	15	5	12	8
2			30	55	10	46	6	26	8
2			17	19	10	19	6	17	8
3				41	10	39	8	23	9
3				18	10	19	7	15	8
4					10	46	6	26	8
4					9	19	6	16	8
5						13	29	19	27
5						12	19	16	18
6							9	29	11
6							8	17	11
7								17	32
7								13	19
8									14
8									12

The upper figure shows the number of primitive variables.
The lower figure shows the number of compound variables

Fig. 6.3 Single-unit realization

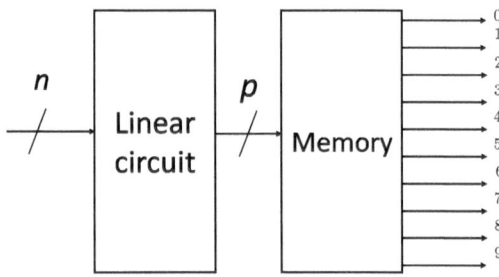

6.4.7 MNIST Two-Class Problem

In the MNIST handwritten digit recognition data set, consider the 784 variable two-class
function f, where the images are binarized by the most significant bit, and

$$f(\vec{a}) = 0 \text{ if } \vec{a} \text{ corresponds to a digit } 0,1,2,3, \text{ or, } 4,$$
$$f(\vec{a}) = 1 \text{ if } \vec{a} \text{ corresponds to a digit } 5,6,7,8, \text{ or, } 9.$$

In this case, $k_0 \simeq k_1 \simeq 2^{15}$. If the function is random, then the number of primitive variables can be reduced to $2 \times 15 - 2 = 30 - 2 = 28$, by Theorem 5.1.2. Algorithm 6.1.1 reduced the primitive variables from $n = 784$ to $p_1 = 35$. This shows that for this function, Theorem 5.1.2 is not applicable.

However, if we use a linear transformation to further reduce the number of variables, then the number of compound variables can be reduced to $p_2 = 24$.

6.4.8 Spam E-Mail Filter (Two-Class Function)

We designed a SPAM E-mail filter with the following specification using the architecture in Fig. 6.1.

- The number of IP address in the white list: 10000.
- The number of IP address in the black list: 10000.
- The number of bits to represent IP address: 128.
- The number of outputs is one. It shows spam ($f = 1$) or ($f = 0$) not.

We also assumed that the bit patterns of IP addresses are random, and that only the E-mails with the IP address in the lists are applied.

When only primitive variables are used: Since $k_0 = k_1 = 10^4$, from Theorem 5.1.1, most functions can be represented with

$$\hat{p}_1 = \lceil \log_2(k_0 k_1) \rceil - 2 = 26 - 2 = 24$$

primitive variables.

When compound variables can be used: The number of distinct vectors in the difference vector set is $N_1 \leq k^2 = 10^8$. From Theorem 6.2.2, any function can be represented with $UB_1 = \lfloor \log_2 N_1 + 1 \rfloor = 26$ variables. Thus, the number of compound variables is at most $\hat{p}_2 = 26$.

A computer simulation using a randomly generated function shows that $p_1 = 25$ (with primitive variables), and $p_2 = 22$ (with compound variables).

Table 6.15 summarizes the experimental results.

6.5 Remarks

This chapter showed a method to reduce the number of compound variables to represent a classification function by a linear transformation. Experimental results using MNIST, fashion MNIST, and spam E-mail filter show the effectiveness of linear transformations. Computational complexity of Algorithms 4.2.1 and 6.1.1 is $O(nk \log k)$ and $O(nN)$, respectively,

Table 6.15 Numbers of variables for large benchmark functions

Data	# inputs	# classes	# primitive variables	# compound variables
	n	m	p_1	p_2
$MNIST$				
$Single-unit$	784	10	37	25
$MNIST$				
$Two-class$	784	2	35	24
$SPAM Mail$				
$Filter$	128	2	25	22

where k is the number of specified minterms, and N is the number of distinct difference vectors. This chapter is based on [2, 14, 15].

6.6   Exercises

6.1 (E) Consider the random classification function $f : B^n \rightarrow \{0, 1, 2, 3\}$ in Example 6.3.4, where $k_0 = k_1 = k_2 = k_3 = 1000$. Estimate the number of compound variables p to represent f using Theorem 6.2.2.

6.2 Consider the 10-unit realization of the MNIST character recognition circuit in Fig. 6.4. The top unit produces a true output if and only if the input image shows the digit 0 (**one-versus-rest classifier**). Estimate the number of compound variables p_0. Assume that the number of images for each digit is 6000.

Fig. 6.4 10-unit realization for MNIST

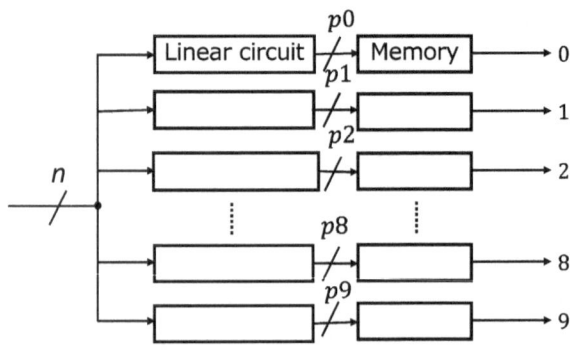

6.3 (E) Consider the English alphabet recognition circuit that appeared in 8.4.5. Assume that the single-unit realization (Fig. 6.3) is used. Estimate the number of compound variables p for the memory. Note that the number of 16-valued variables is 16, and the number of classes (i.e., outputs) is $m = 26$. Also assume that the number of training images for each alphabet is 780.

References

1. Sasao T (2019) Index generation functions. Morgan & Claypool
2. Sasao T (2020) On the minimization of variables to represent partially defined classification functions. ISMVL. Miyazaki, Japan, pp 117–123
3. https://www.countries-ofthe-world.com/all-countries.html
4. https://en.wikipedia.org/wiki/Periodic_table
5. https://str.toyokeizai.net/magazine/shikiho_cd/
6. https://archive.ics.uci.edu/ml/datasets/optical+recognition+of+handwritten+digits
7. Sasao T, Horikawa Y, Iguchi Y (2020) Handwritten digit recognition based on classification functions. ISMVL. Miyazaki, Japan, pp 130–136
8. http://yann.lecun.com/exdb/mnist/
9. https://www.cs.toronto.edu/~kriz/cifar.html
10. https://tech.zalando.com
11. https://atmarkit.itmedia.co.jp/ait/articles/2005/28/news016.html
12. GitHub, Fashion-mnist zalandoresearch, The MIT License (MIT) Copyright c [2017] Zalando SE https://tech.zalando.com
13. Xiao H, Rasul K, Vollgraf R, Fashion-MNIST: a novel image dataset for benchmarking machine learning algorithms. arXiv:1708.07747
14. Sasao T (2020) Reduction methods of variables for large-scale classification functions. In: IWLS, pp 82–87
15. Sasao T (2021) On the number of variables to represent classification functions using linear decompositions. In: SASIMI, pp 29–30. (Virtual workshop)

Data Mining and Machine Learning

7

This chapter introduces data mining and machine learning. It also introduces the concept of generalization, and shows methods to evaluate the performance of supervised machine learning.

7.1 Terminology

First, we review terminology used in data mining and machine learning.

Table 7.1 shows the relations of words used in three different specializations.

The subject of **supervised machine learning** is, given a **training set**, to infer a function that has the **generalization ability**: the ability to predict the outcome for the unseen instances in the **test set**. Note that the training set corresponds to the registered vector table for a classification function.

Memorization just memorizes the training set, and can be performed by just storing the training set into a memory device. On the other hand, **learning** not only memorizes the training set, but also predicts outputs for unknown inputs in the test set. Thus, learning requires the **generalization ability**.

If we generate the model that stores all the examples in the training set exactly, then the model would be too complex. Such a model cannot predict the outcome for unknown inputs in the test set correctly, and has a poor generalization ability.

In this chapter, we consider **explainable machine learning** or **interpretable machine learning** [1]. Such classifiers are often required in medical diagnosis. In medical diagnosis, the model must be simple enough, so that medical doctors can explain the reason for the decision to the patients.

© The Author(s), under exclusive license to Springer Nature Switzerland AG 2024   61
T. Sasao, *Classification Functions for Machine Learning and Data Mining*,
Synthesis Lectures on Digital Circuits & Systems,
https://doi.org/10.1007/978-3-031-35347-5_7

Table 7.1 Terminology in different areas of specialization

Logic design	Geometry	Data mining, machine learning
Minterm	Vertex	Example, instance, sample, object
Implicant	Cube	Rule
Prime implicant	Prime cube	Maximally general rule
SOP	Covering cubes	Covering rule set
Variable	Variable	Feature, attribute

To find a good model, we use **Occam's razor** [2]. That is, to build a model that is consistent with the training set using as simple rules as possible.[1] We also use an SOP to represent the set of rules, and use an SOP minimizer to find a simple rule set.

The subject of data mining is, given a set of examples (minterms), to derive a minimal set of rules that covers the set.

We are interested in deriving simple rules. Simpler rules are more understandable and more efficient to apply [4].

Definition 7.1.1 A **categorical variable** can take one of a limited number of possible values. A **numerical variable** can take on any value within a finite or infinite interval. **Discretization** [5] is to convert numerical variables into categorical ones. Categorical variables include: **nominal variables** and **ordinal variables**. Ordinal variables has an ordering. Numerical variables include: **discrete variables** and **continuous variables**. Discrete variables can be represented by integers. Continuous variables can be represented by real numbers. They include length, weight, temperature, pressure, and time.

Example 7.1.1 Nationality, gender, and color are nominal variables. There is no ordering among them. The roll of a six-sided die is an ordinal variable. Possible outcomes are 1, 2, 3, 4, 5, or 6. Ranking, grading, size of cloth (XS, S, M, L, XL), are ordinal variables.   ∎

In Chap. 8, nominal variables and ordinal variables are used. They should be represented by different binary encodings (See Exercise 8.3). In Sect. 10, we show a method to convert continuous variables into multi-valued variables.

Definition 7.1.2 A set of rules is **complete** if it covers all the examples in each class. A set of rules is **consistent** if each rule covers examples in only one class, and none of the examples in multiple classes.

---

[1] Occam's razor is a rule of thumb, and does not always produce a classifier with high generalization ability [3].

Fig. 7.1 Guess the values for
minterms with ? marks

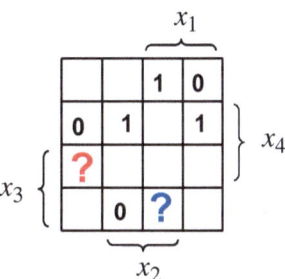

Definition 7.1.3 Two examples are **inconsistent** if they have the same input parts, but belong to different classes.

Definition 7.1.4 **Memorization** is a process to store the data in the training set. **Learning** not only stores the data in the training set, but also predicts the outcome for unseen test set.

Thus, memorization can be performed by just storing the data directly into memory. Learning requires some kind of operation to extract the property of the data.

In the case of neural networks, the weights of neurons are modified to fit the function. At the same time, the interconnections are simplified, and the sum of weights are reduced. Such operation is performed by backpropagation and *Stochastic Gradient Descent*[2] (SGD) [6].

Reduction of variables corresponds to **feature selection** in machine learning and data mining. Reduction of variables using a linear transformation corresponds to **feature reduction**, such as Principal Component Analysis (PCA) [6] in machine learning.

Example 7.1.2 Consider the 4-variable function shown in Fig. 7.1. For the blank cells and cells with question marks, we do not know the function values. We want to guess the function values for the minterms marked by the question marks. Thus, if we guess randomly, the probability of the correct solution is one half for each minterm. However, by analyzing the function carefully, we can guess the correct solution with a higher probability.

After careful analysis of the function, we know that the function is independent of the variable x_3. It can be verified that the map is symmetric with respect of the red horizontal line, as shown in Fig. 7.2. The values for the light blue cells can be guessed when the function is independent of x_3. In this way, the values for two minterms can be guessed.   □

Example 7.1.3 Consider the 4-variable function shown in Fig. 7.3. For the blank cells, we do not know the function values. We want to guess the function values for the minterms with the question marks. In this case, the function depends on all the variables. Thus, we

_____

[2] An optimization method in machine learning.

Fig. 7.2 Guessed values for the
minterms

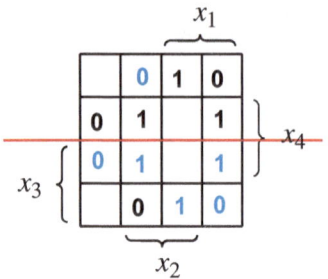

cannot use the previous approach (i.e., detection of a redundant variable). So, we use SOP
minimization to guess the solution.

Figure 7.4 shows the minimized SOP for the function. All the true minterms are covered
by three loops. From this, we can guess the function values for the minterms with question
marks.

Fig. 7.3 Guess the values for
minterms with ? marks

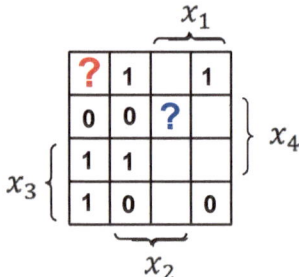

Fig. 7.4 Guess the values for
the minterms by SOP
minimization

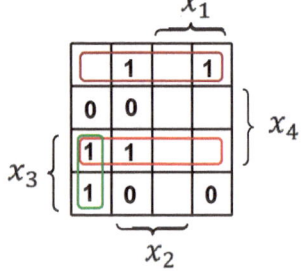

7.2   Comparison with a Tree-Based Classifier

Here, we compare our classifier (**SOP-based classifier**) with a decision tree-based classifier **C4.5** [7].

The goals of the two classifiers are different. The goal of the SOP-based classifier is to
derive a compact set of rules that is consistent with the training set. In this case, the accuracy
for the training set is 1.00. On the other hand, the goal of the C4.5 is to derive a compact
set of rules with high accuracy for both the training set and the test set. In this case, C4.5
seeks the rules with high **generalization ability**. In other words, the goal of the SOP-based
classifier is **logic synthesis** [8] for the training set, while the goal of C4.5 is **approximate
logic synthesis** [9] for the training set.

The SOP-based classifier not only memorizes the training set, but also tries to improve
its generalization ability them by

- Reduction of variables, and
- Assignment of values to *don't care* elements, which is done during SOP minimization.
  (Examples are shown in Figs. 7.4, 8.1, 8.2, 9.2, 9.3, and 9.4).

On the other hand, C4.5 tries to improve its generalization ability by

- Pruning the tree, and
- Limiting the depth of the tree.

7.3   Performance Measures

To evaluate the performance of binary classifiers, we use the **confusion matrix** shown in
Table 7.2, where **TP** is the number of **true positives**, **FP** is the number of **false positives**, **FN**
is the number of **false negatives**, and **TN** is the number of **true negatives**.

First, we define three measures: **Accuracy, Precision,**[3] and **Recall.**[4]

Table 7.2 Confusion matrix

	Predicted	
Actual	Positive	Negative
Positive	TP	FN
Negative	FP	TN

---

[3] Also called positive predictive value [10, 11].
[4] Also called sensitivity or true positive rate [10, 11].

Definition 7.3.1

$$Accuracy = \frac{TP + TN}{TP + FP + FN + TN}$$

$$Precision = \frac{TP}{TP + FP}$$

$$Recall = \frac{TP}{TP + FN}$$

Example 7.3.1 When considering a test to detect infection of COVID-19, it is important to keep in mind that the test is usually not perfect. A positive result indicates that the person has been infected. *Accuracy* is the fraction of all test results that are correct, while *Precision* represents the fraction of truly infected cases where the test results were positive. *Recall*, on the other hand, shows the fraction of truly infected cases detected by the test among all infected cases.

When *Precision* is high, a positive result suggests a high probability of infection. On the other hand, when *Recall* is high, the test detects most positive cases, but it may also classify some negative cases as positive.

When the COVID-19 test has low *Precision*, it can yield false positive results for non-infected individuals, leading to their unnecessary isolation. On the other hand, when the test has low *Recall*, it can produce false negative results for truly infected individuals, which can contribute to the spread of the epidemic. There is a trade-off between *Precision* and *Recall*. (See Exercise 7.2).

Example 7.3.2 One in every 1,000 people has disease A. Test B detects 90% of disease A as positive. It also yields negative results for 95% of those without disease A.

Suppose that you had Test B, and showed a positive result, what is the probability that you have disease A?

You might think that you had the disease A with a high probability, but this intuition is not true.

(Solution)

Assume that a city with 100,000 people. In this case, there are 100 people with disease A, and 99900 people without disease A. Table 7.3 is the confusion matrix. Note that $TP = 90$ and $FP = 4995$. *Precision* is the probability that you are truly positive:

$$Precision = \frac{TP}{TP + FP} = \frac{90}{90 + 4995} \simeq 0.0177.$$

Thus, the probability that you have disease A is not so high as you might think.   □

Definition 7.3.2 In a binary classification, when the number of instances in one class is much smaller than the number of instances in the other class, the set is **imbalanced**.

Table 7.3 Confusion matrix for disease A

	Test positive	Test negative	Total
With disease	TP = 100× 0.9 = 90	FN = 100× 0.1 = 10	100
Without disease	FP = 99,900× 0.05 = 4,995	TN = 99,900× 0.95 = 94,905	99,900
Total	5,085	94,915	100,000

Many practical classification problems have imbalanced class distributions. However, straightforward applications of conventional classifiers to imbalanced sets results in poor solutions.

To show the performance of a classifier for imbalanced data sets, we use **Matthews Correlation Coefficient (MCC)** [11, 12].

$$MCC = \frac{TP \cdot TN - FP \cdot FN}{\sqrt{(TP + FP) \cdot (TP + FN) \cdot (TN + FP) \cdot (TN + FN)}}$$

When $FN = FP = 0$, $MCC = 1.00$, while when $TP = TN = 0$, $MCC = -1.00$. For the classifier that classifies all the instances to *Negative* $(TP = FP = 0)$, or all the instances to *Positive* $(TN = FN = 0)$, MCC is undefined (**UD**). In this case, the denominator of MCC is zero.

Example 7.3.3 Consider *Thoracic* UCI data set [13]. This data is related to the post-operative life expectancy in the lung cancer patients. The total number of instances is 470. Among them, Class 1 (death within one year after surgery) consists of 70 instances, while Class 2 (survival) consists of 400 instances. Note that the number of instances in Class 1 is much smaller than that in Class 2. Thus, the data is imbalanced. The perfect classifier produces the confusion matrix:

$$M_{our} = \begin{bmatrix} 70 & 0 \\ 0 & 400 \end{bmatrix}.$$

On the other hand, J48[5] produced a classifier with the confusion matrix:

$$M_{J48} = \begin{bmatrix} 0 & 70 \\ 0 & 400 \end{bmatrix}.$$

Note that, in the perfect classifier, *Accuracy, Precision, Recall* and *MCC* are all 1.00. On the other hand, in J48, $Accuracy = \frac{400}{400+7} = 0.851$, but $Recall = 0.00$, and *Precision* and *MCC* are undefined.

---

[5] J48 is a Java implementation of C4.5 on **WEKA system** [14].

This shows that for this data, the classifier generated by J48 cannot detect Class 1 patients. Also, it shows that **Accuracy cannot show the performance of the classifier for imbalanced data sets**. □

In machine learning, when the amount of data is insufficient, the performance of the classifier tends to be low. In such a case, artificial data is appended to improve the performance. Such technique is called **data argumentation**. Various methods have been developed to solve imbalanced data sets. Among them, *SMOTE* (Synthetic Minority Over-sampling TechniquE) [15] creates artificial data for minority class [10].

7.4   Evaluation Methods

To evaluate a classifier, one of the following methods is used:

- Use a separate test set, where the training set and the test set have no common elements.
- Use a training set, where the test set is equal to the training set.
- **k-fold Cross Validation**:

  1. Split the training set into k disjoint sets.
  2. Train a model using $(k-1)$-folds as the training set.
  3. Compute the performance measure such as *Accuracy*, by using the remaining part of the data as the test set.
  4. Repeat Step 2–3, for k different test sets.
  5. Compute the average of the k measures.

When the training set is used as the test set, the measure shows the **training accuracy**. When a separate test set is used, the measure shows the real **test accuracy**. Note that the test accuracy is usually lower than the training accuracy.

Example 7.4.1 Figure 7.5 illustrates the 5-fold cross validation. The original training set is split into 5 disjoint sets :$S1, S2, S3, S4, S5$.

1. In the first step, a model is trained using $S2 \cup S3 \cup S4 \cup S5$ as the training set. And, compute *Accuracy 1* using $S1$ as the test set.
2. In the second step, a model is trained using $S1 \cup S3 \cup S4 \cup S5$ as the training set. And, compute *Accuracy 2* using $S2$ as the test set.
3. In the third step, a model is trained using $S1 \cup S2 \cup S4 \cup S5$ as the training set. And, compute *Accuracy 3* using $S3$ as the test set.
4. In the fourth step, a model is trained using $S1 \cup S2 \cup S3 \cup S5$ as the training set. And, compute *Accuracy 4* using $S4$ as the test set.

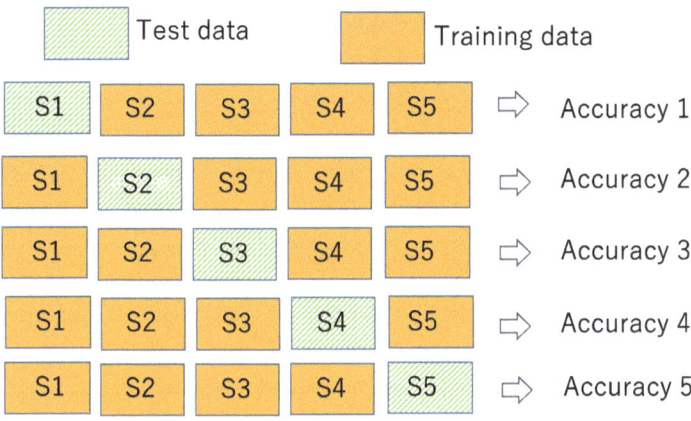

Fig. 7.5 5-fold cross validation

5. In the fifth step, a model is trained using $S1 \cup S2 \cup S3 \cup S4$ as the training set. And, compute *Accuracy 5* using $S5$ as the test set.

The *Accuracy* of the model is the average of the five Accuracies.   □

Example 7.4.2 Consider *Vote* in the UCI data set. This function has 16 variables, and each takes 3 values. The number of instances is 435.

Assume that J48 with the default setting is used. When the training set is used as the test set, we have the confusion matrix:

$$M_{Train} = \begin{bmatrix} 165 & 3 \\ 8 & 259 \end{bmatrix}.$$

In this case, Training Accuracy = $\frac{165+259}{435}$ = 0.9747.

When the 10-fold Cross validation is used, we have the confusion matrix:

$$M_{10-CV} = \begin{bmatrix} 159 & 9 \\ 11 & 256 \end{bmatrix}.$$

In this case, Test Accuracy = $\frac{159+256}{435}$ = 0.954.   ∎

7.5   Accuracy for MNIST

This section compares the test accuracy of various MNIST character recognition circuits.

MNIST data sets consists of the **training set** with 60000 images, and the **test set** with 10000 images. Table 7.4 compares the accuracy of various realizations. ACC1 shows the

Table 7.4 Various realizations of MNIST circuits

Realization	Variable	Architecture	p	ACC1	ACC2
Single-unit	Primitive	Figure 6.3	37.00	0.146	
Single-unit	Compound	Figure 6.3	25.00	0.156	
45-unit	Primitive	Figure 5.5	19.42	0.878	0.901
45-unit	Compound	Figure 5.5	16.86	0.870	0.880
45-unit×4	Primitive	Figure 7.6	15.08	0.907	0.925
45-unit×4	Compound	Figure 7.6	12.77	0.905	0.919

ACC1: Variable minimization only
ACC2: Variable minimization + SOP minimization

test accuracy when the ON set and the OFF set of reduced variables are directly stored in the memory. For unspecified parts, zeros are stored in the memory. ACC2 shows the test accuracy when SOPs of the ON set and the OFF set of reduced variables are simplified and their functions are stored in the memory.

In the **single-unit realization** that appeared in Chap. 6, the accuracies are 0.146 when primitive variables are used, and 0.156 when compound variables are used.

In the **45-unit realizations** that appeared in Chap. 5, test accuracies are improved to 0.878 when primitive variables are used, and 0.870 when compound variables are used. When the SOPs for the ON set and the OFF set are simplified, the test accuracies are further improved to 0.901 in the case of primitive variables, and 0.880 in the case of compound variables. Also, in these cases, the average numbers of primitive and compound variables are reduced to 19.42 and 16.86, respectively.

Next, consider the **45-unit ×4 realization** shown in Fig. 7.6, where the training set is partitioned into four disjoint sets, and each set is implemented by 45-unit, and each counter has $9 \times 4 = 36$ inputs. In this case, the test accuracies are further improved to 0.907 when primitive variables are used, and 0.905 when compound variables are used.[6] When the SOPs for the ON set and the OFF set are simplified, the test accuracies are further improved to 0.925 in the case of primitive variables, and 0.919 in the case of compound variables. In these cases, the average numbers of primitive and compound variables are reduced to 15.08 and 12.77, respectively.

It is interesting to compare the performance with a neural network. In a one-hidden layer fully connected multilayer neural network with 300 hidden units (28×28-300-10), the test accuracy is 0.953 [16]. Thus, the test accuracy of the neural network is higher. Note that in this case, each pixel is represented by 8 bits.

---

[6] In this case, the training accuracy was less than 1.00.

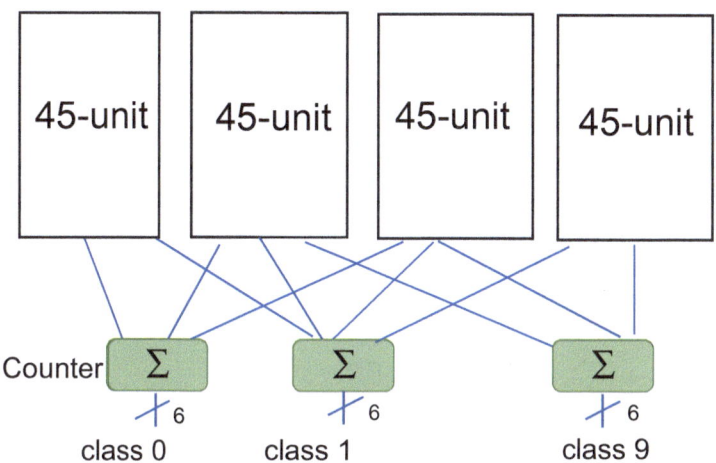

Fig. 7.6 45-unit×4 realization © 2021 IEEE. Reprinted with permission from [17]

7.6   Remarks

This chapter showed the relation among logic design, machine learning, and data mining. Also, it introduced the concept of generalization, and showed methods to evaluate the performance of supervised machine learning.

Zhang et. al [18] showed that **DNNs** (deep neural networks) can memorize random data. Aprit et. al [19] showed that DNNs learn simple patterns first before memorization. DNN optimization is *content-aware*, taking advantage of patterns shared by multiple training examples. Also, they showed that DNN optimization behaviors on real data and random data are different.

Chatterjee [20] considered fan-in limited LUT networks, and analyzed the influence of the number of the layers (levels) and fan-in on the generalization ability of the network. He showed that **LUT networks** can be used for the MNIST [21] problem.

This chapter is based on [22–24].

7.7   Exercises

7.1 *(E)* Assume that the data set consists of $\frac{N}{2}$ true instances, and $\frac{N}{2}$ false instances. Assume that a binary classifier classifies instances randomly with probability 0.5 for true and 0.5 for false, independently of the value of the attributes. That is, the classifier produces outputs by coin tossing. Calculate the *Accuracy*, *Precision*, *Recall* and *MCC* for this classifier.

7.2 To evaluate the performance of a binary classifier, the F_1 **measure** is often used, where

$$F_1 = \frac{2 \cdot Precision \cdot Recall}{Precision + Recall} = \frac{2 \cdot TP}{2 \cdot TP + FP + FN}.$$

F_1 is the **harmonic mean** of Precision and Recall, and it tends to be closer to the smaller of these two measures. Show an example of a confusion matrix, where F_1 cannot show the performance of a binary classifier.

7.3 Consider binary classifiers for an n-variable function:

$$f : \{0, 1\}^n \to \{0, 1\}.$$

Assume that $2k$ vectors are mapped to 0, and $2k$ vectors are mapped to 1. Also, assume that vectors are distinct and random.

1. Estimate the numbers of variables to represent f for a single-unit realization, and a 2-unit realization.
2. Estimate the total amount of memory for a single-unit realization, and a 2-unit realization.
3. Show that the training accuracy can be less than 1.00, in the case of 2-unit realization.

7.4 Compare the amount of memory for MNIST character recognition circuits for the following cases:

1. 45-unit $\times 4$ realization, using compound variables in Fig. 7.6.
2. The neural network that appeared at the end of Sect. 7.5. Assume that each weight of a neuron is represented by an 8-bit number.

References

1. Cano A, Zafra A, Ventura S (2013) An interpretable classification rule mining algorithm. Inf Sci 240:1–20
2. Blumer A, Ehrenfeucht A, Haussler D, Warmuth MK (1987) Occam's razor. Inf Process Lett 24(6):377–380
3. Domingos P (1999) The role of Occam's razor in knowledge discovery. Data Min Knowl Discov 3:409–425
4. Hong SJ (1997) R-MINI: An iterative approach for generating minimal rules from examples. IEEE Trans Knowl Data Eng 9(5):709–717
5. Liu H, Hussain F, Tan CL, Dash M (2002) Discretization: An enabling technique. Data Min Knowl Discov 6:393–423
6. Bishop CM (2006) Pattern recognition and machine learning. Springer

7. Quinlan JR (1993) C4.5: Programs for machine learning. Morgan Kaufmann Publishers, San Mateo, California
8. Hassoun S, Sasao T (eds) (2001) Logic synthesis and verification. Kluwer Academic Publishers
9. Scarabottolo I et al (2020) Approximate logic synthesis: A survey. Proceed IEEE 108(12):2195–2213
10. Branco P, Torgo L, Ribeiro RP (2017) A survey of predictive modeling on imbalanced domains. ACM Comput Surv 49(2):50
11. Chicco D, Jurman G (2020) The advantages of the Matthews correlation coefficient (MCC) over F1 score and accuracy in binary classification evaluation. BMC Genomics 21(6):1–13
12. Boughorbel S, Jarray F, El-Anbari M (2017) Optimal classifier for imbalanced data using Matthews correlation coefficient metric. PLoS ONE 12(6)
13. https://archive.ics.uci.edu/ml/datasets/Thoracic+Surgery+Data
14. https://www.cs.waikato.ac.nz/ml/index.html
15. Chawla NV, Bowyer KW, Hall LO, Kegelmeyer WP (2002) SMOTE: Synthetic minority over-sampling technique. J Artif Int Res 16(1):321–357
16. LeCun Y, Bottou L, Bengio Y, Haffner P (1998) Gradient-based learning applied to document recognition. Proceed IEEE 86(11):2278–2324
17. Sasao T, Horikawa Y, Iguchi Y (2021) A design method for multiclass classifiers. ISMVL, pp 148–153
18. Zhang C, Bengio S, Hardt M, Recht B, Vinyals O (2017) Understanding deep learning requires rethinking generalization. In: Proceedings of the International Conference on Learning Representations, ICLR
19. Arpit D et al (2017) A closer look at memorization in deep networks. ICML-2017, pp 233–242
20. Chatterjee S (2018) Learning and memorization. In: International Conference on Machine Learning (ICML 2018), Stockholm, Sweden, July 10–15, pp 754–762
21. http://yann.lecun.com/exdb/mnist/
22. Sasao T, Horikawa Y, Iguchi Y (2021) Classification functions for handwritten digit recognition. IEICE Trans Inf Syst E104-D(8):1076–1082
23. Sasao T (2022) A method to generate classification rules from examples. ISMVL, May 18–22, pp 176–181
24. Sasao T (2023) Data mining using multi-valued logic minimization. ISMVL, May 22–24

Functions with Multi-valued Inputs

8

In practical applications, variables are often multi-valued. This chapter shows a method to represent functions by an SOP with multi-valued inputs. It also shows that the reduction of variables and simplification of an SOP yields the generalization ability, a power to predict the outcome for unknown inputs. It also presents a method to minimize multi-valued variables. Finally, it shows experimental results using UCI data sets.

8.1   SOP with Multi-valued Inputs

In this part, we review words used in logic synthesis [1].

Definition 8.1.1 A **multi-valued input classification function** is a mapping $f : P_1 \times P_2 \times \cdots \times P_n \to M$ where $P_i = \{1, 2, \ldots, p_i\}$ and $M = \{1, 2, \ldots, m\}$.

Example 8.1.1 An automobile dealer has various models of a car. Each model is classified by the features shown in Table 8.1. Among these models, the following four models are available immediately: (manual, two doors, white), (manual, three doors, blue), (manual, three doors, black), (manual, four doors, red). The following five models can be ordered from the manufacturer, but the customer must wait for several months to get the cars: (automatic, two doors, white), (automatic, two doors, black), (automatic, three doors, blue), (automatic, three doors, black), (automatic, four doors, red). In this table, X_1 shows the transmission type, X_2 shows the number of doors, and X_3 shows the color. If the car model is available immediately, then $F = 1$, can be ordered from the manufacturer, but wait for several months to get the cars, then $F = 2$, otherwise (not available) $F = 3$, where X_1 is two-valued, X_2 is three-valued, and X_3 is four-valued.   ∎

© The Author(s), under exclusive license to Springer Nature Switzerland AG 2024
T. Sasao, *Classification Functions for Machine Learning and Data Mining*,
Synthesis Lectures on Digital Circuits & Systems,
https://doi.org/10.1007/978-3-031-35347-5_8

Table 8.1 Features of automobiles

		1	2	3	4
X_1	Transmission	Manual	Automatic		
X_2	# of doors	Two doors	Three doors	Four doors	
X_3	Color	White	Blue	Red	Black

Definition 8.1.2 Let X be a variable that takes one value in $P = \{1, 2, \ldots, p\}$. Let S be a subset $(S \subseteq P)$ of P. Then, X^S is a **literal** of X. When $X \in S$, $X^S = 1$, and when $X \notin S$, $X^S = 0$. Let $S_i \subseteq P_i$ $(i = 1, 2, \ldots, n)$, then $X_1^{S_1} X_2^{S_2} \cdots X_n^{S_n}$ is a **product**. The product is 1 iff $X_i \in S_i$ for $i = 1, 2, \ldots, n$. $\bigvee_{(S_1, S_2, \ldots, S_n)} X_1^{S_1} X_2^{S_2} \cdots X_n^{S_n}$ is a **sum-of-products expression** (**SOP**). An SOP is 1 iff at least one product is 1. When $S_i = P_i$, $X_i^{S_i} = 1$ and the product is independent of X_i. In this case, literal $X_i^{P_i}$ is redundant and can be deleted. A product corresponds to a **cube**. When $|S_i| = 1$ $(i = 1, 2, \ldots, n)$, a product corresponds to an element of the domain. This product is a **minterm of** f. When $S_i = P_i$ $(i = 1, 2, \ldots, n)$, the product corresponds to the constant 1. This product corresponds to a **universal cube**. **Cube size** is the total number of **vertices** contained in the cube. The **size of the universal cube** is $\Pi_{i=1}^n p_i$. It is often called the **size of the universe**.

When $p_i = 2$ $(i = 1, 2, \ldots, n)$, a function is a two-valued logic function. An arbitrary multi-valued input classification function is represented by an SOP. Many SOPs exist that represent the same function. Among them, the one with the minimum number of products is the **minimum SOP** (**MSOP**).

Example 8.1.2 Consider the automobile dealer in the previous example. Let $P_1 = \{1, 2\}$, $P_2 = \{1, 2, 3\}$, $P_3 = \{1, 2, 3, 4\}$. Let F_1 be the set of automobiles in the inventory. Let F_2 be the set of available automobiles, but not in the inventory. Let F_3 be the set of automobiles that are unavailable. Then, we have

$$F_1 = \{(1, 1, 1), (1, 2, 2), (1, 2, 4), (1, 3, 3)\},$$
$$F_2 = \{(2, 1, 1), (2, 1, 4), (2, 2, 2), (2, 2, 4), (2, 3, 3)\}, \text{ and}$$
$$F_3 = \{\text{Other combinations}\}.$$

The size of the universal cube is $p_1 \times p_2 \times p_3 = 2 \times 3 \times 4 = 24$. ∎

Example 8.1.3 Consider the set of automobiles in the previous example. It represents a multi-valued input classification function,

$$\begin{aligned}
F_1 &= X_1^{\{1\}} X_2^{\{1\}} X_3^{\{1\}} \vee X_1^{\{1\}} X_2^{\{2\}} X_3^{\{2\}} \vee X_1^{\{1\}} X_2^{\{2\}} X_3^{\{4\}} \\
&\quad \vee X_1^{\{1\}} X_2^{\{3\}} X_3^{\{3\}} \\
&= X_1^{\{1\}} X_2^{\{1\}} X_3^{\{1\}} \vee X_1^{\{1\}} X_2^{\{2\}} X_3^{\{2,4\}} \vee X_1^{\{1\}} X_2^{\{3\}} X_3^{\{3\}}
\end{aligned}$$

and

$$F_2 = X_1^{\{2\}} X_2^{\{1\}} X_3^{\{1\}} \vee X_1^{\{2\}} X_2^{\{1\}} X_3^{\{4\}} \vee X_1^{\{2\}} X_2^{\{2\}} X_3^{\{2\}} \vee$$
$$X_1^{\{2\}} X_2^{\{2\}} X_3^{\{4\}} \vee X_1^{\{2\}} X_2^{\{3\}} X_3^{\{3\}}$$
$$= X_1^{\{2\}} X_2^{\{1\}} X_3^{\{1,4\}} \vee X_1^{\{2\}} X_2^{\{2\}} X_3^{\{2,4\}} \vee X_1^{\{2\}} X_2^{\{3\}} X_3^{\{3\}}. \qquad \blacksquare$$

Let the domain of a multi-valued input two-valued output function be $P_1 \times P_2 \times \cdots \times P_n$. In this case, the product $c = X_1^{S_1} X_2^{S_2} \cdots X_n^{S_n}$, where $S_i \subseteq P_i$ corresponds to a **cube** in an n-dimensional hyper-cube. The **bit representation** (positional cube notation) of a cube c is the concatenation of binary numbers showing the cube. $c = \pi_1 - \pi_2 - \cdots - \pi_n$, where $\pi_i = (\xi_0 \xi_1 \cdots \xi_{p_i-1})$, such that

$$\xi_j = \begin{cases} 1 & (\text{when } j \in S_i), \\ 0 & (\text{when } j \notin S_i). \end{cases}$$

The bit representation of an SOP is an **array**. An array is a set of cubes.

Example 8.1.4 The last SOP for F_2 of the previous example is represented by an array:

$$\begin{matrix} X_1 \ X_2 \quad X_3 \\ 12 \ -123 \ -1234 \end{matrix}$$
$$\begin{bmatrix} 01 \ -100 \ -1001 \\ 01 \ -010 \ -0101 \\ 01 \ -001 \ -0010 \end{bmatrix}$$

Note that a variable with all 1's in the bit representation can be omitted in the SOP. $\blacksquare$

Definition 8.1.3 An SOP is called a **disjoint sum-of-products expression** (DSOP), if all the products are mutually disjoint.

Theorem 8.1.1 ([1]) *To represent an* $(n = 2r)$*-variable logic function*

$$x_1 x_2 \vee x_3 x_4 \vee \cdots \vee x_{n-1} x_n,$$

an SOP requires r products, while a DSOP requires $2^r - 1$ *products.*

8.2   Minimization of Multi-valued Variables

In this part, we show a method to minimize the number of variables in the support set.

Definition 8.2.1 Let $f : D \to M$ be a classification function with n variables, where $M = \{1, 2, \ldots, m\}$. If there exist $\vec{a} = (a_1, a_2, \ldots, a_n), \vec{b} = (b_1, b_2, \ldots, b_n) \in D$. such that $a_j = b_j$ for every $j \in \{1, 2, \ldots, n\} \setminus \{i\}$, $a_i \neq b_i$, and $f(\vec{a}) \neq f(\vec{b})$, then f is said to **depend on** the i-th variable. When a function f depends on x_i, then x_i is said to be **essential**.

This is an extension of binary case (Definition 3.1.1) to multi-valued case.

Theorem 8.2.1 *If a function f depends on X_i, then any support set of f contains i.*

(Proof) If a support set does not contain i, then there are no vectors $\vec{a}$ and $\vec{b}$ such that $f(\vec{a}) \neq f(\vec{b})$ and $\vec{e}_i = \vec{a} \oplus \vec{b}$. This means that f is independent of X_i.   □

When a function f depends on x_i, then a literal for x_i appears in any SOP for f. This is an extension of the binary case. An example of a binary case appeared in Example 3.1.1.

Example 8.2.1 Consider the function shown in Table 8.2. It depends on X_4, since

$$\vec{a}_2 = (1, 3, 2, 3) \in F_1 \quad and$$
$$\vec{c}_2 = (1, 3, 2, 1) \in F_3.$$

Thus, X_4 is essential.   ∎

The following algorithm is an extension of two-valued cases (Algorithm 3.2.1) to multi-valued cases.

Table 8.2 Registered vector table

	X_1	X_2	X_3	X_4	f	
$\vec{a}_1$	2	3	3	2	1	F_1
$\vec{a}_2$	1	3	2	3	1	
$\vec{b}_1$	3	2	2	1	2	F_2
$\vec{b}_2$	3	1	2	1	2	
$\vec{c}_1$	2	3	2	1	3	F_3
$\vec{c}_2$	1	3	2	1	3	

Algorithm 8.2.1 (Minimization of Multi-Valued Variables)

1. For each pair $(\vec{a}, \vec{b})$ such that $\vec{a} \in F_\alpha$ and $\vec{b} \in F_\beta$, $(\alpha \neq \beta)$ make a clause

$$C(\vec{a}, \vec{b}) = z_1 \vee z_2 \vee \cdots \vee z_n,$$

   where

$$z_j = \begin{cases} 0 & \text{if } a_j = b_j \\ y_i & \text{if } a_j \neq b_j. \end{cases}$$

2. For all the pairs $(\vec{a}, \vec{b})$ in $\vec{a} \in F_\alpha$ and $\vec{b} \in F_\beta$, make the product of the clauses.

$$R = \bigwedge_{(\vec{a}, \vec{b})} C(\vec{a}, \vec{b}).$$

3. Covert the expression of R into an SOP, and simplify it. A product with the fewest literals corresponds to a minimum support set.

Example 8.2.2 Consider the function in Table 8.2.

1. The set of clauses are:

$$\begin{array}{ll} C(\vec{a}_1, \vec{b}_1) = y_1 \vee y_2 \vee y_3 \vee y_4, & C(\vec{a}_1, \vec{b}_2) = y_2 \vee y_3 \vee y_4, \\ C(\vec{a}_1, \vec{c}_1) = y_3 \vee y_4, & C(\vec{a}_1, \vec{c}_2) = y_1 \vee y_3 \vee y_4, \\ C(\vec{a}_2, \vec{b}_1) = y_1 \vee y_2 \vee y_4, & C(\vec{a}_2, \vec{b}_2) = y_1 \vee y_2 \vee y_4, \\ C(\vec{a}_2, \vec{c}_1) = y_1 \vee y_4, & C(\vec{a}_2, \vec{c}_2) = y_4, \\ C(\vec{b}_1, \vec{c}_1) = y_1 \vee y_2, & C(\vec{b}_1, \vec{c}_2) = y_1 \vee y_2, \\ C(\vec{b}_2, \vec{c}_1) = y_1 \vee y_2, & C(\vec{b}_2, \vec{c}_2) = y_1 \vee y_2, \end{array}$$

2. Forming the product of all the clauses yields:

$$R = (y_1 \vee y_2 \vee y_3 \vee y_4)(y_2 \vee y_3 \vee y_4)(y_3 \vee y_4)$$
$$(y_1 \vee y_3 \vee y_4)(y_1 \vee y_2 \vee y_4)(y_1 \vee y_4)y_4(y_1 \vee y_2).$$

3. The simplified expression is $R = (y_1 \vee y_2)y_4 = y_1 y_4 \vee y_2 y_4$.

The minimum support set is $\{1, 4\}$ and $\{2, 4\}$. ∎

8.3 SOP Minimization and Generalization Ability

In data mining, each row in Table 8.2 corresponds to an example, an instance, or a sample.
    The goal in data mining is to find a simple representation of the sample set. Minimization of the number of rules corresponds to minimization of the number of products in an SOP.

Table 8.3 Reduced registered vector table

X_2	X_4	f
3	2	1
3	3	1
2	1	2
1	1	2
3	1	3
3	1	3

In logic design, specifications are often given by the ON sets and the DC (*don't care*) sets. On the other hand, in data mining, specifications are given by examples. Combinations not shown by the examples correspond to the DC set.

Example 8.3.1 Consider the function in Table 8.3. By a minimization of an SOP of multi-valued inputs, we have the following:

$$F_1 = X_4^{\{2,3\}}, \quad F_2 = X_2^{\{1,2\}} \cdot X_4^{\{1\}}, \quad F_3 = X_2^{\{3\}} \cdot X_4^{\{1\}}.$$

When English is used to represent the function, we have:

1. If $(X_4 = 2 \ OR \ 3)$, then $f = 1$, else
2. if $(X_2 = 1 \ OR \ 2)$ AND $(X_4 = 1)$, then $f = 2$, else
3. if $(X_2 = 3)$ AND $(X_4 = 1)$, then $f = 3$.

For the function in Table 8.2, the output values for only 6 input vectors are specified, while the number of possible input vectors is $3^4 = 81$. Thus, the values for other $81 - 6 = 75$ vectors are undefined. On the other hand, in Table 8.3, each row shows $3^2 = 9$ vectors, and the last two rows are the same. Thus, Table 8.3 shows $5 \times 9 = 45$ combinations. For example, for the vector $(X_1, X_2, X_3, X_4) = (1, 3, 1, 2)$, the output is undefined in Table 8.2. However, in Table 8.3, the output is specified to $f = 1$. Figure 8.1 shows the map of the rules. In this case, value 1 are assigned to four blank cells. Note that the function values for these four vectors are undefined in Table 8.3. The SOP minimizer assigned the value 1 to these cells. Thus, the rules have the **generalization ability** for unseen input vectors. However, the SOP minimizer may produce the map in Fig. 8.2. In this case, value 2 are assigned to four blank cells.

Note that the SOP minimizer never assigns value 3 to these blank cells.   ∎

This example showed that the generalization ability is given by (1) reduction of variables, and (2) simplification of the SOP.

Fig. 8.1 Value 1 are assigned

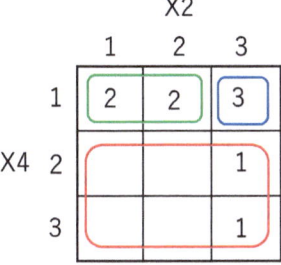

Fig. 8.2 Value 2 are assigned

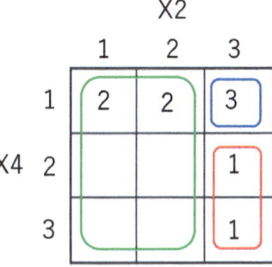

Example 8.3.2 Consider the 6-variable function $f(x_1, x_2, \ldots, x_6)$ in Table 3.1. This function can be represented with only three variables (x_1, x_2, x_4).

$$\mathcal{F}_1 = \bar{x}_1 \bar{x}_4 \vee x_2 \bar{x}_4.$$

$$\mathcal{F}_2 = x_4 \vee x_1 \bar{x}_2.$$

Figure 8.3 shows the map of the simplified SOP. Originally, values for $k_1 + k_2 = 3 + 3 = 6$ minterms are specified. After the variable minimization, $6 \times 2^3 = 48$ minterms are specified. Thus, the function values for $48 - 8 = 42$ minterms are guessed by the variable reduction. The SOP minimizer found the property that f is independent of x_3, x_5 and x_6. After the SOP minimization, values for $8 \times 8 = 64$ minterms are specified. Thus, the values for $64 - 6 = 58$ minterms of f are guessed.   ∎

Fig. 8.3 Map for the 3-variable function

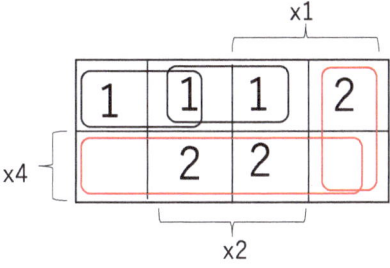

8.4   Experiments with UCI Data Sets

To show the effectiveness of the approach, we derived rules for selected benchmark functions from the UCI (University of California, Irvine) machine learning repository [2].

    Table 8.4 summarizes the experimental results. The first column shows the name of the function. The second column shows the number of original variables: n. The third column shows the number of instances: k. The fourth column shows the number of classes: m. The fifth column shows the number of variables after variable minimization: n_e. The sixth column shows the number of rules after SOP minimization: r_e. Rules are generated for m classes. For example, in the case of *Monks* ($m = 2$), rules for both F_1 and F_2 were generated. However, in many cases, only one of two is sufficient. The seventh column shows the **rule complexity** or PLA measure: $\alpha = n_e \cdot r_e$. The eighth column shows n_t, the number of variables obtained by **J48**.[1] The ninth column shows r_t, the number of rules obtained by J48. The tenth column shows the rule complexity of J48: $\beta = n_t \cdot r_t$. The eleventh column shows the accuracy for J48.[2] The last four columns show the same things for **JRIP**. To minimize the number of variables, Algorithm 8.2.1 was used, while to minimize SOPs with multi-valued inputs, heuristic algorithms in [1, 5, 6] were used.

8.4.1   Breast Cancer

This function classifies the patients into two classes: (1) malignant, and (2) benign. The original dataset consists of 699 instances. We removed 16 incomplete instances (i.e., instances with missing entries), and six conflicting instances. Each variable takes 10 values. In the original data, the variables were continuous, but they were discretized to 10-values.

8.4.2   Chess3196

This function shows whether white can win or not for $k = 3196$ starting positions. We assume that each player plays optimally. Variables take 2, 3, or 4 values.

8.4.3   Connect-4

This function classifies the positions of the game into three outcomes: (1) player X win, (2) player Y wins, or (3) draw. We assume that each player plays optimally. Each variable takes 3 values: empty, X, or Y.

---

[1] J48 is a Java implementation of C4.5 on **WEKA system** [3]. **C4.5** is a standard decision tree classifier [4].

[2] As for the test set, the training set was used. Thus, Acc shows the **training accuracy**.

8.4.4 Dermatology

This function classifies the patients into six classes. 10 conflicting instances were removed from the original data. Variables take 2, 3, 4, or 76 values.

8.4.5 Letter Recognition

This function classifies the image data of English alphabets into 26 classes. Each variable takes 16 values.

8.4.6 Monks

This function shows the relation of input variables. There are 6 variables: two of them take 2 values; three of them take 3 values; and one of them takes 4 values. The number of instances is $k = 2^2 \times 3^3 \times 4 = 432$. Note that this function is **totally defined**. X_1, X_2, and X_5 are essential. The minimal sets of rules are:

$$F_1 = X_5^{\{1\}} \vee X_1^{\{1\}} X_2^{\{1\}} \vee X_1^{\{2\}} X_2^{\{2\}} \vee X_1^{\{3\}} X_2^{\{3\}}.$$
$$F_2 = X_1^{\{1\}} X_2^{\{2,3\}} X_5^{\{2,3,4\}} \vee X_1^{\{2\}} X_2^{\{1,3\}} X_5^{\{2,3,4\}}$$
$$\vee X_1^{\{3\}} X_2^{\{1,2\}} X_5^{\{2,3,4\}}.$$

An alternative expression for F_1 is,

$$F_1 = (X_1 == X_2) \vee (X_5 == 1).$$

8.4.7 Mushroom

This function classifies the mushrooms into two classes: (1) edible, and (2) not edible. Originally, it contained 8124 instances, but 2480 incomplete instances were removed. Note that [6] shows a solution with five rules which depends on 8 variables, while our method produced a solution with five rules which depends on only 3 variables.

An SOP representing poisonous mushrooms is

$$Poison = X_5^{\{2,3,4,5,7\}} \vee X_{21}^{\{0,1,2,3,5\}} X_{22}^{\{1,5\}} \vee X_{21}^{\{1,4\}} X_{22}^{\{0,2,5\}},$$

where

- X_5 denotes odor: almond (0), anise (1), creosote (2), fishy (3), foul (4), musty (5), none (6), pungent (7), spicy (8);

- X_{21} denotes population: abundant (0), clustered (1), numerous (2), scattered (3), several (4), solitary (5); and
- X_{22} denotes habitat: grasses (0), leaves (1), meadows (2), paths (3), urban (4), waste (5), woods (6).

This SOP covers the all the poisonous mushrooms in the training set, while none of edible ones. However, it also covers mushrooms not in the training set. To minimize the number of variables, we used Algorithm 8.2.1. To simplify the multi-valued SOP, we used MINI2 [1], which is an improved version of MINI [5].

8.4.8 Promoter

This function classifies the gene sequences (DNA) into two classes: (1) positive or (2) negative. Each variable takes 4 values: (1) A, (2) G, (3) T, and (4) C.

8.4.9 Splice

This function classifies the gene sequences (DNA) into three classes: (1) donor, (2) acceptor, (3) neither. Originally, it contained 3191 instances, but 16 ambiguous instances were removed. Each variable takes 4 values: (1) A, (2) G, (3) T, and (4) C.

8.4.10 Teaching Assistant Evaluation

This function classifies the performance of teaching assistants into three classes (1) low, (2) medium, and (3) high. Originally, it contained 151 instances, but 8 conflicting instances were removed. Variables take 2, 23 or 42 values.

8.4.11 Tic-Tac-Toe

This function classifies the patterns of the marking in a 3×3 grid into two classes: (1) player X wins, and (2) player X does not win. We assume that each player plays optimally. Each variable takes 3 values.

8.4.12 Vote

This function classifies the members of U.S. House of Representatives into two classes: (1) democrat, and (2) republican. Each variable takes 3 values.

8.4.13 Zoo

This function classifies 101 animals into 7 classes: (1) Mammals, (2) Birds, (3) Reptiles, (4) Fish, (5) Amphibians, (6) Insects, (7) Others.

Among 16 variables, X_{13} shows the number of legs and takes 6 values. Other variables are binary. Two variables X_6 and X_{13} are essential. There are seven minimal sets consisting of five variables. Among them, when X_4, X_6, X_9, X_{12}, and X_{13} are used to represent the function, we have the following rules:

$$F_1 = X_4^{\{1\}}.$$
$$F_2 = X_4^{\{0\}} X_{13}^{\{2\}}.$$
$$F_3 = X_4^{\{0\}} X_6^{\{0\}} X_{13}^{\{4\}} \vee X_9^{\{1\}} X_{12}^{\{0\}} X_{13}^{\{0\}}.$$
$$F_4 = X_4^{\{0\}} X_{12}^{\{1\}}.$$
$$F_5 = X_4^{\{0\}} X_6^{\{1\}} X_9^{\{1\}} X_{13}^{\{4\}}.$$
$$F_6 = X_6^{\{0\}} X_{13}^{\{6\}}.$$
$$F_7 = X_9^{\{0\}} X_{13}^{\{0,8\}} \vee X_6^{\{1\}} X_{13}^{\{4,5,6,8\}}.$$

Each variable represents the following:

- $X_4 = 1$ iff it gives milk,
- $X_6 = 1$ iff it is aquatic,
- $X_9 = 1$ iff it has a backbone,
- $X_{12} = 1$ iff it has fins,
- X_{13} shows the number of legs, and takes a value in $\{0, 2, 4, 5, 6, 8\}$.

When English is used to represent the classes, we have[3]

1. It is a **mammal** if it gives milk, else
2. it is a **bird** if it gives no milk and has two legs, else

---

[3] Reptiles include a seasnake, which is aquatic and has no leg. Others include a scorpion, which is not aquatic and has eight legs; a crab, which is aquatic and has four legs; a starfish, which is aquatic and has five legs; a lobster, which is aquatic and has six legs; and an octopus, which is aquatic and has eight legs.

3. it is a **reptile** if it gives no milk, is not aquatic and has four legs, or if it has a backbone but no fins and has no leg, else
4. it is a **fish** if it gives no milk and has fins, else
5. it is an **amphibian** if it gives no milk, is aquatic, has a backbone, and has four legs, else
6. it is an **insect** if it is not aquatic and has six legs, else
7. it is **another animal** if it has no backbone and has no or eight legs, or it is aquatic and has four, five, six or eight legs.

8.5   Experiments with Imbalanced Data Sets

In a binary classification, when the number of instances in one class is much larger than the number of instances in the other class, the set is **imbalanced** [8, 9]. Many practical classification problems have imbalanced class distributions.

In this part, we consider experimental results for imbalanced data sets. Table 8.5 shows the results. The meanings of the columns are the same as that of Table 8.4.[4] In this part, we use the following data sets.

8.5.1   Hepatitis

This function classifies the patients into two classes: (1) death and (2) survival. Incomplete instances were removed. The number of instances in Class 1 and Class 2 are 13 and 67, respectively. Variables take 2, 5, 6, 7, or 11 values.

8.5.2   Thoracic

This data is related to the post-operative life expectancy in the lung cancer patients. The number of instances is 470. Among them, Class 1 (death within one year after surgery) consists of 70 instances, while Class 2 (survival) consists of 400 instances. The number of variables is $n = 16$: three of them are numerical, one takes 3, one takes 4, one takes 7 values, and ten of them take 2 values.

---

[4] As for the test set, the training set was used. Thus, Acc shows the training accuracy.

Table 8.4 Result for standard benchmark sets. © 2023 IEEE. Reprinted, with permission, from [7]

| | Orig. var | Inst. | | Our classifier | | | J48 | | | | JRIP | | | |
	n	k	m	Var n_e	Rule r_e	Comp. α	Var n_t	Rule r_t	Comp. β	Acc. %	Var n_t	Rule r_t	Comp. γ	Acc. %
Breast cancer	9	677	2	4	15	60	7	12	84	99.0	7	6	42	98.5
Chess3196	36	3196	2	29	41	1189	22	30	660	99.7	22	17	374	99.6
Connect-4	42	67557	3	34	8356	284104	42	3871	162582	86.9	33	82	2706	75.5
Dermatology	34	358	6	6	53	318	7	8	56	98.0	10	8	80	96.9
Letter recog	16	20000	26	11	1394	15334	16	1226	19616	96.3	16	428	6848	93.6
Monks	6	432	2	3	7	21	1	2	2	75.0	1	2	2	75.0
Mushrooms	22	5644	2	3	5	15	6	8	48	100.0	6	6	36	100.0
Promoter	57	106	2	4	10	40	7	10	70	98.1	7	4	28	93.4
Splice	60	3174	3	10	544	5440	41	101	4141	98.4	29	13	377	96.0
Teaching assis	5	143	3	3	18	54	5	35	175	88.8	3	4	12	58.0
Tic Tac Toe	9	958	2	8	62	496	9	49	441	93.9	9	11	99	98.7
Vote	16	435	2	9	23	207	5	7	35	97.5	1	2	2	94.5
Zoo	16	101	7	5	9	45	8	9	72	99.0	8	7	56	94.1

Table 8.5 Results for imbalanced data sets. © 2023 IEEE. Reprinted, with permission, from [7]

	Orig.		Class	Our classifier			J48				JRIP			
	Var.	Inst.		Var.	Rule	Comp.	Var.	Rule	Comp.	MCC	Var.	Rule	Comp.	MCC
	n	k	m	n_e	r_e	α	n_t	r_t	β		n_t	r_t	γ	
Hepatitis	19	80	2	4	8	32	3	4	12	0.703	2	2	4	0.689
Thoracic	16	470	2	4	28	112	0	1	0	UD	0	1	0	UD

8.6   Observations

8.6.1   Analysis of Table 8.4

For *Monks*, J48 and JRIP produced poor solutions, i.e., the accuracy was 75%, although the size of their trees was small. Among 15 benchmark functions, only this function was totally defined. That is, the values of this function are specified for all possible input vectors.

For *Connect-4*, J48 and JRIP produced poor solutions, i.e., the accuracy was 86.9%, and 75.5%, respectively, while our classifier produced 100% accuracy. However, the rule complexities for J48 and JRIP were smaller.

For *Mushrooms*, three classifiers produced solutions with 100% accuracy, while our classifier produced the simplest solution.

In general, J48 produced solutions with higher accuracy than JRIP. Our classifier always produced solutions with 100% accuracy.

8.6.2   Analysis of Table 8.5

Here, we analyze the results for imbalanced data sets.

For *hepatitis*, our classifier produced 8 rules with four variables, and with MCC = 1.00, while J48 produced a decision tree with three variables, and four rules, and with MCC = 0.793. JRIP produced two rules with two variables and MCC = 0.689.

For *Thoracic*, our classifier produced 28 rules with four variables and MCC = 1.00, while J48 produced a decision tree with only one leaf, and classified all the data into Class 2 (survival). In this case, MCC = UD. Thus, J48 is useless for this data set. JRIP produced the same solution.

8.7   Remarks

This chapter showed a method to simplify multi-valued SOPs to reduce the number of rules. This method produces sets of rules with 100% test accuracy for a given set of examples. It also showed that the reduction of variables and the simplification of the SOP give the generalization ability.

Then, it compared the number of rules with C4.5 (J48) and JRIP. This method is useful for data mining, especially for bio-medical applications, since the obtained rules are simple and explainable. In bio-medical applications, the numbers of the samples are relatively small, and the data are often imbalanced, while rules with very high accuracy are required.

In this chapter, an m-valued variable was represented by m bits. A direct representation of a model using multi-valued variables is easy to interpret. However, it is also possible to use m binary variables to represent an m-valued variable (**one-hot encoding**), or to use $\lceil \log_2 m \rceil$ binary variables to represent an m-valued variable (**minimum-length encoding**). In these cases, the number of variables can be reduced [10].

Most n-variable functions require $O(2^n)$ products in their SOPs, however many functions that appear in the UCI data sets for machine learning can be represented by SOPs with small numbers of products. Exceptions are Connect-4, letter recognition, and splice, etc. This is similar to the case of MCNC benchmark functions for logic synthesis. They can also be represented with SOPs with a small numbers of products. Exceptions are parity functions and multipliers.

Chapter 9 considers a class of functions whose SOPs have a small numbers of products. This chapter is based on [1, 7].

8.8   Exercises

8.1 Consider the Mushroom data set in Sect. 8.4.7. After removing incomplete data, the function represents
$$P_1 \times P_2 \times \cdots \times P_{22} \to B,$$
where $P_i = \{0, 1, \ldots, p_i - 1\}$ and

$$
\begin{aligned}
&p_1 = 6, \quad p_2 = 4, \quad p_3 = 10, \quad p_4 = 2, \quad p_5 = 8, \quad p_6 = 4, \\
&p_7 = 3, \quad p_8 = 2, \quad p_9 = 12, \quad p_{10} = 2, \quad p_{11} = 6, \quad p_{12} = 4, \\
&p_{13} = 4, \quad p_{14} = 9, \quad p_{15} = 9, \quad p_{16} = 1, \quad p_{17} = 2, \quad p_{18} = 3, \\
&p_{19} = 5, \quad p_{20} = 8, \quad p_{21} = 6, \quad p_{22} = 7.
\end{aligned}
$$

The simplified expression for the poisonous mushrooms is

$$F = X_5^{\{2,3,4,5,7\}} \vee X_{21}^{\{0,1,2,3,5\}} X_{22}^{\{1,5\}} \vee X_{21}^{\{1,4\}} X_{22}^{\{0,2,5\}}.$$

Calculate the number of combinations covered by F.

8.2 (E) Consider the 6-variable function $f(x_1, x_2, \ldots, x_6)$ in Fig. 5.1. This function can be represented with only three variables (x_2, x_3, x_6). Simplify the SOP for this function, and count the minterms whose values are guessed by the variable reduction and by SOP minimization.

Table 8.6 One-hot code and thermometer code

n	One-hot code	Thermometer code
0	100000	00000
1	010000	10000
2	001000	11000
3	000100	11100
4	000010	11110
5	000001	11111

8.3 To represent an ordinal variable, **thermometer encoding** is suitable, while to represent a nominal variable, **one-hot encoding** is suitable. Explain why.

For example, numbers {0, 1, 2, 3, 4, 5} can be represented by one-hot code and by thermometer code as shown in Table 8.6.

References

1. Sasao T (1999) Switching theory for logic synthesis. Kluwer Academic Publishers
2. https://archive.ics.uci.edu/ml/datasets.php
3. https://www.cs.waikato.ac.nz/ml/index.html
4. Wu X, Kumar V, Quinlan JR et al (2008) Top 10 algorithms in data mining. Knowl Inf Syst 14:1–37
5. Hong SJ, Cain RG, Ostapko DL (1974) MINI: A heuristic approach for logic minimization. IBM J Res Develop 443–458
6. Hong SJ (1997) R-MINI: An iterative approach for generating minimal rules from examples. IEEE Trans Knowl Data Eng 9(5):709–717
7. Sasao T (2023) Data mining using multi-valued logic minimization. ISMVL, May 22–24
8. Branco P, Torgo L, Ribeiro RP (2017) A survey of predictive modeling on imbalanced domains. ACM Comput Surv 49(2):50. Article 31
9. Krawczyk B (2016) Learning from imbalanced data: open challenges and future directions. Progr Artif Intell 5:221–232
10. Sasao T, Butler JT (2021) Linear decompositions for multi-valued input classification functions. ISMVL. May 2021, pp 13–18

Easily Reconstructable Functions

9

This chapter shows that simplification of SOPs gives the generalization ability. We show this in three steps. First, various classes of totally defined logic functions (SOPs) were generated. Second, minterms of the function are randomly selected to generate partially defined functions. And, third, from the partially defined functions, original functions are reconstructed by SOP minimization. We consider the Achilles heel functions, majority functions, monotone increasing cascade functions, functions generated from random SOPs, and monotone increasing random SOPs. As for machine learning methods, Naive Bayes, multi-level perceptron, support vector machine, JRIP, J48, and random forest are considered in addition to SOP minimization.

9.1   Reconstruction of Functions by SOP Minimization

SOP minimization finds the simplest SOP that is consistent with the given partially defined function. This technique is often used to design economical digital circuits. However, it can be also used for **reconstruction of a function**.

Definition 9.1.1 A **minimum SOP** has the fewest products, and the number of literals is also minimal.

Lemma 9.1.1 ([1]) *Any* **minimum SOP** *can be represented as a sum of* **prime implicants (PIs).**

Definition 9.1.2 The **SOP degree** of f is the maximum number of literals in a product of a minimum SOP across all products.

© The Author(s), under exclusive license to Springer Nature Switzerland AG 2024   91
T. Sasao, *Classification Functions for Machine Learning and Data Mining*,
Synthesis Lectures on Digital Circuits & Systems,
https://doi.org/10.1007/978-3-031-35347-5_9

Table 9.1 Partially defined function

	x_1	x_2	x_3	x_4	f
ON	1	1	0	0	1
	1	1	1	0	1
	1	1	1	1	1
OFF	0	0	0	0	0
	0	1	0	0	0
	0	1	0	1	0
	1	0	0	0	0
	1	0	1	0	0

Fig. 9.1 Partially defined function f

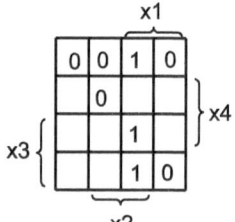

Example 9.1.1 Consider the two-class function f whose ON and OFF sets are shown in Table 9.1. Suppose that Table 9.1 is the training set. Predict the output for the function for the input $\vec{a} = (x_1, x_2, x_3, x_4) = (1, 1, 0, 1)$. Since the vector $\vec{a}$ is not contained in the training set, nobody knows the output for this input.

Figure 9.1 shows the map for f, where blank cells denote *don't cares*. If we perform an SOP minimization using the map in Fig. 9.2, we have the simplified expression for f:

$$\mathcal{F} = x_1 x_2.$$

Although the value of $f(\vec{a})$ is undefined, if the value is 1, and if the values of f for other blank cells are 0, then the rule for f becomes simpler. **Occam's razor** [2] recommends to use simple rules. So, we assume that the function value for $\vec{a}$ to be 1.

Next, predict the output value for the input $\vec{b} = (0, 0, 0, 1)$. In this case, the output is assumed to be zero, since this makes the rule simpler. Such an operation corresponds to a **generalization** in supervised machine learning.   ∎

Fig. 9.2 Simplified SOP for
the function $f : \mathcal{F}$

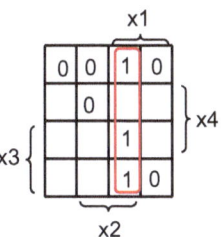

Lemma 9.1.2 *The SOP degrees of functions are*

- *1 for n-variable OR function.*
- *n for n-variable AND and* **parity function.**
- *(r + 1) for (2r + 1)-variable* **majority function.**
- *2 for (2r)-variable* **Achilles heel function** *[1]:*

$$Ach2(r) = x_1 y_1 \vee x_2 y_2 \vee \cdots \vee x_r y_r.$$

Example 9.1.2 Consider the Achilles heel function with 6 variables:

$$Ach2(3) = x_1 y_1 \vee x_2 y_2 \vee x_3 y_3.$$

The complement of $Ach2(3)$ is

$$
\begin{aligned}
\overline{Ach2(3)} &= (\bar{x}_1 \vee \bar{y}_1)(\bar{x}_2 \vee \bar{y}_2)(\bar{x}_3 \vee \bar{y}_3) \\
&= \bar{x}_1 \bar{x}_2 \bar{x}_3 \vee \bar{x}_1 \bar{x}_2 \bar{y}_3 \vee \bar{x}_1 \bar{y}_2 \bar{x}_3 \vee \bar{x}_1 \bar{y}_2 \bar{y}_3 \vee \\
&\quad \bar{y}_1 \bar{x}_2 \bar{x}_3 \vee \bar{y}_1 \bar{x}_2 \bar{y}_3 \vee \bar{y}_1 \bar{y}_2 \bar{x}_3 \vee \bar{y}_1 \bar{y}_2 \bar{y}_3.
\end{aligned}
$$

The last SOP is minimum, and the maximum number of literals is three. So, the SOP degree
of $\overline{Ach2(3)}$ is three.   ∎

Example 9.1.3 Consider the two-class function f in Table 9.2, where the OFF and the ON
sets are shown. Note that the ON set consists of two vectors: $\{\vec{a}_1, \vec{a}_2\}$, while the OFF set
consists of two vectors: $\{\vec{b}_1, \vec{b}_2\}$.

The minimal SOPs for f are $\mathcal{F}_1 = x_1 x_4 \vee \bar{x}_1 \bar{x}_4$, and $\mathcal{F}_2 = \bar{x}_2 \vee x_3$. The maps for these
SOPs are shown in Figs. 9.3 and 9.4. The minimal SOPs for $\bar{f}$ are $\mathcal{F}_3 = x_1 \bar{x}_4 \vee \bar{x}_1 x_4$, and
$\mathcal{F}_4 = x_2 \bar{x}_3$. Thus, $\mathcal{F}_4$ is the exact minimum for $\bar{f}$. The maps of these SOPs are shown in
Figs. 9.5 and 9.6. For $\bar{f}$, the maximal number of literals in a product is two. Note that these
SOPs depend on only two variables.   ∎

Table 9.2 Partially defined function

		x_1	x_2	x_3	x_4	f
ON	$\vec{a}_1$	1	0	0	1	1
	$\vec{a}_2$	0	1	1	0	1
OFF	$\vec{b}_1$	1	1	0	0	0
	$\vec{b}_2$	0	1	0	1	0

Fig. 9.3 SOP for $f : \mathcal{F}_1$

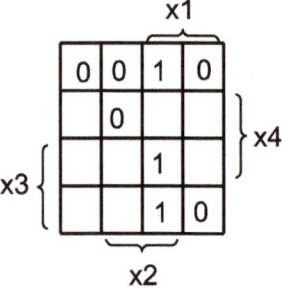

Fig. 9.4 SOP for $f : \mathcal{F}_2$

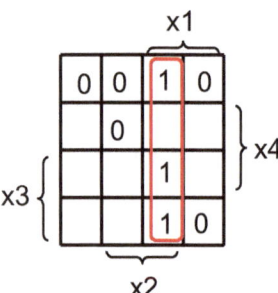

Fig. 9.5 SOP for $\bar{f} : \mathcal{F}_3$

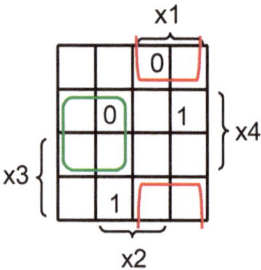

Definition 9.1.3 Let $\vec{a}$ and $\vec{b}$ be elements of B^n. If f satisfies $f(\vec{a}) \geq f(\vec{b})$,[1] for any vectors such that $\vec{a} \geq \vec{b}$, then f is a **monotone increasing function**.

---

[1] Let $\vec{a} = (a_1, a_2, \ldots, a_n)$ and $\vec{b} = (b_1, b_2, \ldots, b_n)$. Then, $\vec{a} \geq \vec{b}$ iff $a_i \geq b_i$ for all i.

Fig. 9.6 SOP for $\tilde{f} : \mathcal{F}_4$

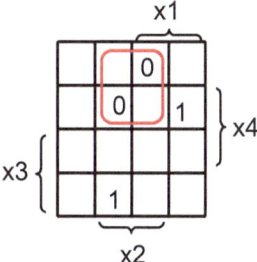

Theorem 9.1.1 ([1]) *Let f be a monotone increasing function of n variables. Then,*

(1) All the prime implicants of f are **essential prime implicants**.
(2) The minimal SOP for f is **unique**.

9.2 Experiments

9.2.1 SOP Minimizer for Machine Learning

In this experiment, a modified version of **MINI** [3, 4] SOP minimizer was used. It produces a ternary output: (f_1, f_0), where $(1,0)$ shows the positive class, $(0,1)$ shows the negative class, and $(0,0)$ shows the unknown class. The following algorithm shows the outline:

Algorithm 9.2.1 (MINI13)

1. From the ON and the OFF sets, generate the DC set by $DC = \overline{ON \cup OFF}$.
2. Simplify the SOP for the ON set, and the SOP for the OFF set using DC, independently. The SOP for the ON set covers all the true minterms and possibly some *don't care* minterms. And, the SOP for the OFF set covers all the false minterms and possibly some *don't care* minterms.
3. Count the products in the simplified SOPs. Let f_1 and f_0 be the totally defined functions for the simplified SOPs for the ON and the OFF sets, respectively.
4. If the simplified SOP for f_1 has fewer products, then replace the SOP for f_0 by the simplified SOP for $f_0 \overline{f_1}$. Otherwise, replace the simplified SOP for f_1 by the simplified SOP for $f_1 \overline{f_0}$.

The last step is used to make the resulting SOPs **disjoint**. Also, it **arguments the data** of the minority class to improve MCC.

Fig. 9.7 Maps for the ON set
and the OFF set

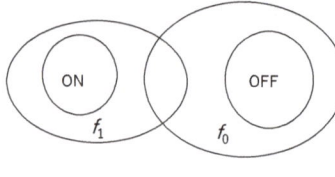

Fig. 9.8 Maps for f_1 and $f_0 \bar{f_1}$

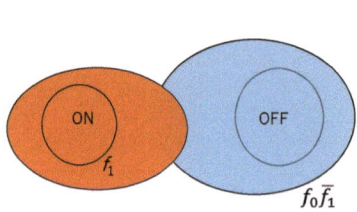

Example 9.2.1 Let Fig. 9.7 be the map for the ON set and the OFF set. Let f_1 be the function for the simplified SOP for the ON set, and f_0 be the function for the simplified SOP for the OFF set. Assume that f_1 and f_0 have common elements, and the number of products in the SOP for f_1 is smaller than that for f_0.

The red area in Fig. 9.8 shows the map for f_1, while the blue area in Fig. 9.8 shows the map for $f_0 \bar{f_1}$. The number of the products in the SOP for $f_0 \bar{f_1}$ is often larger than that for f_0.   ∎

9.2.2   Achilles Heel Functions

Example 9.2.2 Consider the following Achilles heel functions:

$$Ach2(6) = x_1 y_1 \vee x_2 y_2 \vee x_3 y_3 \vee x_4 y_4 \vee x_5 y_5 \vee x_6 y_6.$$
$$Ach3(4) = x_1 y_1 z_1 \vee x_2 y_2 z_2 \vee x_3 y_3 z_3 \vee x_4 y_4 z_4.$$
$$Ach4(3) = x_1 y_1 z_1 w_1 \vee x_2 y_2 z_2 w_2 \vee x_3 y_3 z_3 w_4.$$
$$Ach6(2) = x_1 y_1 z_1 w_1 u_1 v_1 \vee x_2 y_2 z_2 w_2 u_2 v_2.$$

These functions have 12 variables. Thus, the total number of input combinations is $2^{12} = 4096$. Let $|f|$ be the number of minterms in the ON sets for f. Then, $|Ach2(6)| = 3367$, $|Ach3(4)| = 1695$, $|Ach4(3)| = 721$, and $|Ach6(2)| = 127$.

We generated partially defined functions with different numbers of selected minterms, and tried to reconstruct the original functions by the SOP minimizer. The results are shown in Table 9.3. In the table, ○ shows that the SOP minimizer successfully reconstructed the original function, while × shows that the SOP minimizer failed to reconstruct the original function. This result shows that $Ach3(4)$ and $Ach4(3)$ are easier to reconstruct, while

Table 9.3 Reconstruction of Achilles heel functions ($n = 12$). © 2023 IEEE. Reprinted, with permission, from [5]

	# of minterms			# of products		SOP
	100	200	400	f	Not (f)	Degree
Ach2(6)	×	○	○	6	64	2
Ach3(4)	○	○	○	4	81	3
Ach4(3)	○	○	○	3	64	4
Ach6(2)	×	×	○	2	36	6

*Ach*6(2) is harder to reconstruct. Also, with the increase of the number of selected minterms, the reconstruction of functions became easier. ∎

9.2.3 Other Monotone Increasing Functions

A **symmetric threshold function** with n variables and threshold T is defined as

$$Th(n, T)(x_1, x_2, \ldots, x_n) = 1 \Leftrightarrow \sum_{i=1}^{n} x_i \geq T.$$

It is a monotone increasing function.

Th(8,4) is an 8-variable symmetric threshold function, where the threshold is four. SOPs for Th(8,4) has $\binom{8}{4} = 70$ prime implicants, while $\overline{Th(8, 4)}$ has $\binom{8}{5} = 56$ prime implicants.

Table 9.4 shows the experimental results. SOP minimizations could not reconstruct Th(n,T). For these functions, the number of the prime implicants is large.

A **monotone cascade function** (MCF) can be realized by a cascade of two-input logic cells, where OR cells and AND cells are connected alternately. For example,

Table 9.4 Reconstruction of other monotone increasing functions ($n = 8$). © 2023 IEEE. Reprinted, with permission, from [5]

	# of minterms				# of products		SOP
	60	120	160	200	f	Not(f)	Degree
Th(8,2)	×	×	×	×	28	8	2
Th(8,4)	×	×	×	×	70	56	4
MCF(8)	×	×	○	○	5	4	5

Table 9.5 Reconstruction of random functions ($n = 8$). © 2023 IEEE. Reprinted, with permission, from [5]

	# of minterms				# of products		SOP
	80	100	120	160	f	Not(f)	Degree
SOP(8,4,3)	○	○	○	○	4	11	3
SOP(8,5,3)	×	×	×	○	5	15	3
SOP(8,6,3)	×	×	○	○	6	11	3
MoSOP(8,4,3)	×	○	○	○	4	13	3
MoSOP(8,5,3)	×	○	○	○	5	17	3
MoSOP(8,6,3)	×	○	○	○	6	15	3

$$MCF(8) = (((x_1 \vee x_2)x_3 \vee x_4)x_5 \vee x_6)x_7 \vee x_8.$$

The numbers of the prime implicants of MCF(n) is $\lfloor \frac{n}{2} \rfloor + 1$, while that of the complement of MCF(n) is $\lceil \frac{n}{2} \rceil$.

Table 9.4 also shows the results. Reconstruction of the MCF(n) function from the function whose minterms were randomly selected was easier than that of the Th(n,T) function.

9.2.4 Functions Generated by Random SOPs

To specify the complexity of SOPs, we use the following parameters: n denotes the number of variables; m denotes the number of products; and k denotes the number of literals.[2]

SOP(n, m, k) is a randomly generated non-monotone SOP with n variables, m products, and k literals in each product.

MoSOP(n, m, k) is a randomly generated monotone SOP with n variables, m products, and k positive literals in each product.

Table 9.5 compares the experimental results for 8-variable functions. In the table, entries for MoSOPs have more ○ marks than those for SOPs. Thus, MoSOPs are easier to reconstruct than SOPs.

9.3 Evaluation of Methods

Example 9.3.1 Consider the function shown in Fig. 9.9. It is the Achilles heel function with 4 variables:

$$Ach2(2) = x_1 x_2 \vee x_3 x_4.$$

---

[2] Only in this section, m denotes the number of products, and k denotes the number of literals in a product.

Fig. 9.9 Achilles heel function f

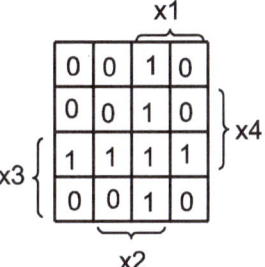

Fig. 9.10 Function after logic minimization of $\hat{f}$

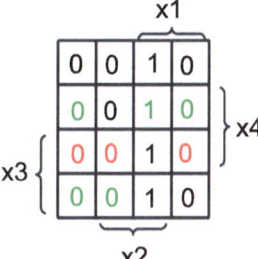

Consider the function shown in Fig. 9.1, where eight minterms are randomly selected. The blank cells are unknown (*don't cares*).

Figure 9.10 shows the minimized function. The colored cells are predicted by the SOP minimizer. Among them, red cells are incorrectly predicted, while green cells are correctly predicted. From Fig. 9.10, we have the confusion matrix:

$$\begin{bmatrix} TP & FN \\ FP & TN \end{bmatrix} = \begin{bmatrix} 4 & 3 \\ 0 & 9 \end{bmatrix}.$$

From this, we have Accuracy $= 0.8125$, and MCC $= 0.6547$.   ∎

In Sect. 9.2, an SOP minimizer reconstructed the original functions from the partially defined functions whose minterms were randomly selected. In Tables 9.3, 9.4 and 9.5, instances with ○ show that the reconstructions are perfect. That is FP $=$ FN $= 0$, and Accuracy $=$ MCC $= 1.00$.

These experiments show that SOP minimization produces the **generalization ability**.

Example 9.3.2 Consider the case of SOP(8,5,3) with $s = 120$ selected minterms. It is shown in the first group of columns in Table 9.7. The SOP is

$$\bar{x}_6\bar{x}_7\bar{x}_8 \vee \bar{x}_4x_6\bar{x}_7 \vee x_3x_4\bar{x}_5 \vee x_2\bar{x}_3\bar{x}_4 \vee \bar{x}_1x_2\bar{x}_5.$$

Note that the number of true minterms is 124.

Consider the function whose 120 minterms of SOP(8,5,3) are randomly selected. When the SOPs for the ON set and the OFF set are simplified independently, the SOP for the ON (positive) set requires 5 products, while the SOP for the OFF (negative) set requires 8 products. So, the SOP for the ON set is kept as it is, while the SOP for the OFF set is slimmed. In this case, the number of products for the OFF set increase to 11 by the slim operation. The confusion matrix is

$$
\begin{bmatrix} TP & FN \\ FP & TN \end{bmatrix} = \begin{bmatrix} 124 & 0 \\ 11 & 118 \end{bmatrix}.
$$

Note that $TP + FN + FP + TN = 253 < 2^8$. Thus, for $256 - 253 = 3$ input vectors, this classifier produces **unknown** outputs. In this case,

$$
Accuracy = \frac{TP + TN}{TP + FN + FP + TN} = 0.957,
$$

$$
Precision = \frac{TP}{TP + FP} = 0.918,
$$

and

$$
Recall = \frac{TP}{TP + FN} = 1.000. \qquad\blacksquare
$$

9.4   Comparison with Other Classifiers

9.4.1   Performance of the Proposed Method

Here, we analyze ten cases with × marks in Tables 9.3, 9.4 and 9.5. The last row headed with MINI13 in Tables 9.6 and 9.7 show the Accuracy(ACC) and MCC obtained by the proposed method. Bold figures show the largest MCC. These data show that the SOP minimizer predicted unknown data well. The Accuracy is, in many cases, acceptable. However, MCC for Ach2(6) and Ach6(2) is lower. Note that for Ach2(6), $|ON| = 3369$ and $|OFF| = 727$, while for Ach6(2), $|ON| = 127$ and $|OFF| = 3969$. In other words, these data sets are imbalanced. Imbalanced data sets are known to be hard to reconstruct [6].

9.4.2   Performance of Other Methods

We investigated the performance of the following classifiers in **WEKA system** [7, 8].

- **Bayes** is a statistical learning algorithm based on Bayes' theorem. It is also called Naive Bayes method, and assumes that variables are independent.

Table 9.6 Accuracy and MCC for various classifiers (Two-valued inputs). © 2023 IEEE. Reprinted, with permission, from [5]

| | Ach2(6) | | Ach6(2) | | Th(8,2) | | Th(8,4) | | MCF(8) | |
| | 100 minterms | | 200 minterms | | 200 minterms | | 200 minterms | | 120 minterms | |
	ACC	MCC	ACC	MCC	ACC	MCC	ACC	MCC	ACC	MCC
Bayes	0.851	0.373	0.972	0.345	0.968	0.328	0.972	0.942	0.953	0.896
MLP	0.863	0.502	0.961	0.396	1.000	**1.000**	1.000	**1.000**	0.992	0.983
SMO	0.864	0.457	0.969	UD	0.996	0.941	1.000	**1.000**	0.961	0.918
JRIP	0.832	0.402	0.969	**0.494**	0.972	0.700	0.867	0.724	0.980	0.956
J48	0.840	0.407	0.969	UD	0.965	0.453	0.816	0.599	0.988	0.974
RF	0.870	0.479	0.971	0.246	0.996	0.941	0.965	0.924	0.996	**0.991**
MINI13	0.874	**0.561**	0.965	0.479	0.980	0.760	0.945	0.887	0.996	**0.991**

UD denotes undefined

Table 9.7 Accuracy and MCC for various classifiers (Two-valued inputs). © 2023 IEEE. Reprinted, with permission, from [5]

| | SOP(8,5,3) | | SOP(8,6,3) | | MoSOP(8,4,3) | | MoSOP(8,5,3) | | MoSOP(8,6,3) | |
| | 120 minterms | | 100 minterms | | 80 minterms | | 80 minterms | | 80 minterms | |
	ACC	MCC	ACC	MCC	ACC	MCC	ACC	MCC	ACC	MCC
Bayes	0.738	0.476	0.730	0.441	0.809	0.575	0.812	0.615	0.801	0.597
MLP	0.770	0.544	0.883	0.759	0.832	0.638	0.859	0.707	0.879	0.755
SMO	0.730	0.463	0.746	0.479	0.797	0.546	0.844	0.674	0.840	0.674
JRIP	0.773	0.547	0.895	0.784	0.875	0.726	1.000	**1.000**	0.938	**0.874**
J48	0.766	0.535	0.832	0.679	0.859	0.689	0.812	0.609	0.906	0.813
RF	0.832	0.664	0.906	0.807	0.910	**0.804**	0.887	0.764	0.902	0.803
MINI13	0.957	**0.917**	0.984	**0.966**	0.877	0.737	0.879	0.753	0.892	0.783

- **MLP** (a Multi-Layer Perceptron) is a feed-forward artificial neural network.
- **SMO** is an extension of a support vector machine using a sequential minimal optimization algorithm [9].
- **JRIP** is a rule learner based on the RIPPER (Repeatedly Incremental Pruning to Produce Error Reduction) algorithm [10].
- **J48** is a decision tree classifier, and is a Java implementation of C4.5 algorithm [11].
- **Random Forest** (**RF**) is an **ensemble classifier** that consists of many decision trees.

The parameters for classifiers were set to default values of WEKA. As for the test set, we used the set of all input combinations,[3] i.e., 2^n vectors, since we know the correct values for all possible input combinations. Tables 9.6 and 9.7 also compare the performance of other methods: Bayes, MLP, SMO, JRIP, J48, and RF.

- The bold numbers show the highest MCCs.
- The method that produced rules with the highest MCC also produced rules with the highest *Accuracy*.
- Bayes produced the lowest MCCs for five functions.
- MLP produced the highest MCCs for Th(8,2) and Th(8,4).
- SMO produced the highest MCCs for Th(8,4). This result is reasonable, since Th(8,4) can be represented by a simple **threshold gate**.
- JRIP produced the highest MCCs for three functions.
- RF produced the highest MCCs for two functions.
- MINI13 produced the highest MCCs for four functions.
- For Ach6(2), two algorithms SMO and J48 produced MCC with UD. In these cases, the algorithms classified all the data into the negative class. Thus, TP = FP = 0, and the values of MCC and *Precision* are undefined (UD), while *Recall* is 0.00. This function is imbalanced, and is hard to reconstruct.

In addition to the *Accuracy* and *MCC*, we have to consider the complexity of the models (rules). In general, MLP, J48 and RF are too complex to analyze, while SOPs generated by JRIP and MINI13 are relatively easy to analyze.

9.5 Extension to Multi-valued Input Functions

In this part, we extend the theory to multi-valued input functions. For simplicity, we consider the data sets for the functions that are generated by random SOPs:

$$f : \{0, 1, 2, 3\}^5 \to \{0, 1\}.$$

We assume that each SOP has m products and k literals, and each literal has one of the following forms:

$$X^{\{3\}}, X^{\{2,3\}}, X^{\{1,2,3\}}, X^{\{0,1,2,3\}}.$$

This implies that the functions are monotone increasing. We did similar experiments to the two-valued input cases. Table 9.8 shows the experimental results.

The first column shows the function name: $MVSOP(n, m, k)$, where n is the number of variables, m is the number of products, and k is the number of literals in a product. Note that

---

[3] This is different from a common method to measure the accuracy, since generation of all possible input combinations is usually impractical.

Table 9.8 Accuracy and MCC for random SOPs (Four-valued inputs:MINI13). © 2023 IEEE. Reprinted, with permission, from [5]

Function	100 minterms		150 minterms		200 minterms		300 minterms		400 minterms		600 minterms	
	ACC	MCC	ACC	MCC	ACC	MCC	ACC	MCC	ACC	MCC	ACC	MCC
MVSOP(5,3,2)	0.973	**0.933**	1.000	**1.000**	1.000	**1.000**	1.000	**1.000**	1.000	**1.000**	1.000	**1.000**
MVSOP(5,4,2)	0.968	**0.918**	0.972	**0.929**	1.000	**1.000**	1.000	**1.000**	1.000	**1.000**	1.000	**1.000**
MVSOP(5,6,2)	0.978	**0.932**	0.951	**0.868**	0.970	0.909	0.979	0.936	0.984	0.953	1.000	**1.000**
MVSOP(5,3,3)	0.911	0.813	0.968	0.933	1.000	**1.000**	1.000	**1.000**	1.000	**1.000**	1.000	**1.000**
MVSOP(5,4,3)	0.949	**0.902**	0.985	**0.971**	0.997	**0.994**	0.997	0.994	0.997	0.994	0.997	0.994
MVSOP(5,6,3)	0.939	**0.878**	0.873	**0.762**	0.957	0.913	0.996	**0.992**	0.990	0.980	0.992	0.984

Table 9.9 Accuracy and MCC for random SOPs (Four-valued inputs:JRIP). © 2023 IEEE. Reprinted, with permission, from [5]

Function	100 minterms		150 minterms		200 minterms		300 minterms		400 minterms		600 minterms	
	ACC	MCC	ACC	MCC	ACC	MCC	ACC	MCC	ACC	MCC	ACC	MCC
MVSOP(5,3,2)	0.941	0.873	0.973	0.933	0.988	0.971	0.988	0.971	1.000	**1.000**	1.000	**1.000**
MVSOP(5,4,2)	0.910	0.787	0.957	0.893	0.988	0.972	1.000	**1.000**	1.000	**1.000**	1.000	**1.000**
MVSOP(5,6,2)	0.885	0.610	0.947	0.846	0.990	.970	0.990	**0.970**	0.994	**0.983**	0.998	0.994
MVSOP(5,3,3)	0.953	**0.900**	1.000	**1.000**	1.000	**1.000**	1.000	**1.000**	1.000	**1.000**	1.000	**1.000**
MVSOP(5,4,3)	0.921	0.850	0.974	0.948	0.996	0.992	1.000	**1.000**	1.000	**1.000**	1.000	**1.000**
MVSOP(5,6,3)	0.889	0.783	0.936	0.871	0.959	**0.918**	0.957	0.914	0.992	**0.984**	0.996	**0.992**

the total number of input combinations for each function is $4^5 = 1024$. Bold figures show the largest MCC.

The first row of Table 9.8 shows that to reconstruct $MVSOP(5, 3, 2)$, 100 selected minterms were not sufficient, but with 150 selected minterms, MINI13 could reconstruct the original function. With the increase of the selected minterms, the *Accuracy* and *MCC* tend to increase. Also, with the increase of the products (m) and literals (k), the reconstruction tends to be more difficult.

When the number of products m and the number of literals k in each product are small, less than 10% of the input combinations are sufficient to reconstruct more than 90% of the entries in the truth table of the original functions. Thus, SOP minimizations produced the generalization ability in the case of multi-valued inputs.

To compare with an existing method, we did similar experiments using the JRIP program, which is a rule-based method for machine learning [10]. Table 9.9 shows the results. Roughly speaking, MINI13 produced competitive solutions to JRIP.

9.6   Remarks

This chapter showed that simplification of an SOP produces the generalization ability. Experimental results showed that the following functions are easily reconstructable:

- Functions with a small number of products in their SOPs.
- Functions with a small number of literals in each products in their SOPs.
- Monotone functions.

However, imbalanced data sets are harder to reconstruct. Machine learning using SOP simplification is promising for imbalanced data.

Learning of Boolean functions was considered in the papers [12–14]. Reconstruction of partially defined monotone increasing logic functions was considered in [15]. It analyzed the complexity of computing. Learning of logic functions using support vector machines was considered in [16]. It compared its performance with C4.5 and naive Bayes classifiers. Experimental results of a rule-based classifier for various benchmark functions were also shown in [17]. Reconstructions of logic functions in an IWLS contest appear in [18].

This chapter is based on [5].

9.7   Exercises

9.1 Show that the number of true minterms for the Achilles heel function:

$$f = x_1x_2 \vee x_3x_4 \vee \cdots \vee x_{n-1}x_n,$$

where $n = 2m$ is

$$\eta(n) = \sum_{i=1}^{m}(-1)^{i+1}\binom{m}{i}2^{n-2i}.$$

Compute the value of $\eta(n)$ for $n = 12$.

9.2 By using J48 in WEKA system [7], derive the rules to classify the function in Table 3.1. Use the default parameters of WEKA.

References

1. Sasao T (1999) Switching theory for logic synthesis. Kluwer Academic Publishers
2. Domingos P (1999) The role of Occam's razor in knowledge discovery. Data Min Knowl Discov 3:409–425

3. Hong SJ, Cain RG, Ostapko DL (1974) MINI: a heuristic approach for logic minimization. IBM J Res Dev 443–458

4. Hong SJ (1997) R-MINI: an iterative approach for generating minimal rules from examples. IEEE Trans Knowl Data Eng 9(5):709–717

5. Sasao T (2023) Easily reconstructable logic functions. ISMVL, May 22–24

6. Krawczyk B (2016) Learning from imbalanced data: open challenges and future directions. Prog Artif Intell 5:221–232

7. https://www.cs.waikato.ac.nz/ml/index.html

8. Witten I, Frank E, Hall M (2016) Data mining: practical machine learning tools and techniques. Morgan Kaufmann

9. Platt JC (1998) Fast training of support vector machines using sequential minimal optimization. In: Schoelkopf B, Burges C, Smola A (eds) Advances in kernel methods - support vector learning. MIT Press

10. Cohen WW (1995) Fast effective rule induction. In: Twelfth international conference on machine learning, pp 115–123

11. Quinlan JR (1993) C4.5: programs for machine learning. Morgan Kaufmann Publishers, San Mateo, California

12. Blum A, Burcht C, Langford J (1998) On learning monotone Boolean functions. In: Proceedings of the symposium on foundations of computer science, FOCS-1998, pp 408–415

13. Natarajan BK (1987) On learning Boolean functions. In: Proceedings of the ACM symposium on theory of computing, (STOC-1987), pp 296–304

14. Rivest RL (1987) Learning decision lists. Mach Learn 2:229–246

15. Muselli M, Ferrari E (2011) Coupling logical analysis of data and shadow clustering for partially defined positive Boolean function reconstruction. IEEE Trans Knowl Data Eng 23(1):37–50

16. Sadohara K (2002) On a capacity control using Boolean kernels for the learning of Boolean functions. In: Proceedings of the IEEE international conference on data mining, pp 410–417

17. Ibrahim MH, Hacibeyoglu M (2020) A novel switching function approach for data mining classification problems. Soft Comput 24(13):4941–4957

18. Rai et al S (2021) Logic synthesis meets machine learning: trading exactness for generalization. DATE2021, pp 1026–1031

Functions with Continuous Variables

10

Classification functions in life science and engineering often have continuous variables. Such variables show length, width, weight, pressure, and temperature, etc. This chapter shows a method to convert continuous variables into discrete ones. Experimental results using UCI data sets are presented.

10.1   Generation of Rules from Examples

To show the method, we use an example to detect people afflicted with liver cirrhosis [1].

Example 10.1.1   Table 10.1 shows the results of blood tests for 10 healthy people (**negative**) and 10 people afflicted with liver cirrhosis (**positive**). Three tests are used to detect liver cirrhosis. ZTT (Zinc sulfate Turbidity Test) indicates chronic hepatitis or liver cirrhosis when the value is greater than 12.0. ALT (Alanine Aminotransferase Test) indicates liver damage. An ALT value greater than 30 indicates some problem. ALB (Albumin blood test) measures the amount of albumin. An ALB value lower than 4.0 indicates liver disease or infection. In the diagnosis column of Table 10.1, 1 shows negative (not afflicted with cirrhosis), while 2 shows positive (afflicted with liver cirrhosis).

   To derive simple rules to detect liver cirrhosis, decision trees are often used. First, ZTT is used to partition the people into two classes. If ZTT is greater than 12.15, then the person is positive. (Strictly speaking, the person with ID 12 is negative, and the other 8 people are positive.) If ZTT is less than or equal to 12.15, then ALB is used to partition the people into two classes. If ALB is greater than 3.75, then the person is negative. Otherwise, the person is positive. Figure 10.1 shows the decision tree to find liver cirrhosis.   ■

Table 10.1 Result of blood test. © 2022 IEEE. Reprinted with permission from [2]

ID	ZTT	ALT	ALB	Diagnosis
1	10.6	25	4.9	1
2	11.2	33	4.9	1
3	11.5	18	4.0	1
4	11.6	22	5.5	1
5	11.6	25	4.4	1
6	11.7	28	4.4	1
7	11.7	37	4.7	1
8	11.7	30	3.7	2
9	11.9	30	4.8	1
10	11.9	35	3.6	2
11	12.1	30	3.8	1
12	12.2	32	4.3	1
13	12.2	34	4.1	2
14	12.2	35	4.4	2
15	12.4	23	3.5	2
16	12.5	37	3.5	2
17	12.6	32	3.3	2
18	12.8	41	3.9	2
19	12.9	28	3.7	2
20	13.3	36	4.1	2

Diagnosis: 1: Negative, 2: Positive (Liver Cirrhosis)

Example 10.1.2 Given a decision tree, the set of rules can be derived from the path from the root node to a leaf node. Figure 10.1 produces three rules:

Rule 1 : If $(ZTT \leq 12.15)$ and $(ALB \leq 3.75)$, then positive.
Rule 2 : If $(ZTT \leq 12.15)$ and $(ALB > 3.75)$, then negative.
Rule 3 : If $(ZTT > 12.15)$, then positive.

This set of rules is complete, but produces a wrong result for the patient with $ID = 12$. ∎

Reduction of the number of rules corresponds to SOP minimization. In logic design, specifications are often given by the ON sets and the DC (*don't care*) sets. On the other hand, in data mining, specifications are given by examples. The numbers of examples in each classes range in the thousands to millions. The size of the universe in data mining can be larger than that of a typical logic circuit. For example, *Dermatology* that appears in Sect. 10.3, has 34 variables. After the domain reduction, 21 variables take 4 values, 8 variables

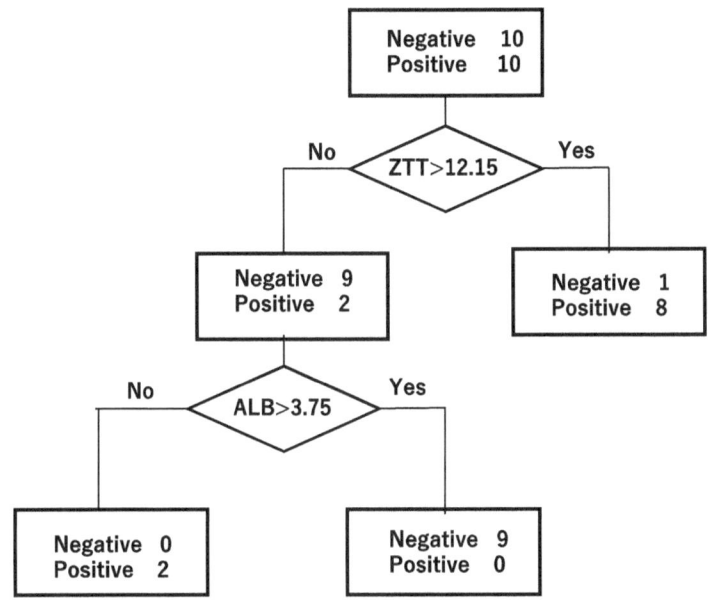

Fig. 10.1 Decision tree to find liver cirrhosis © 2022 IEEE. Reprinted with permission from [2]

take 3 values, 4 variables take 2 values, and one variable takes 59 values. Thus, the size of the universe is

$$4^{21} \times 3^8 \times 2^4 \times 59^1 \simeq 2.723 \times 10^{19}.$$

This is approximately equal to the number of assignments of values to the variables of a binary logic function with $n = 65$ variables, since

$$2^{65} \simeq 3.6 \times 10^{19}.$$

10.2 Algorithms and Examples

10.2.1 Discretization

Algorithm 10.2.1 *(Discretization)*

1. Let $\vec{R}(i) = (R_1(i), R_2(i), \ldots, R_n(i))$ be the data for the i-th example, where $i = 1, 2, \ldots, k$, and let $Class(i)$ be the class of the i-th example.
2. For $j = 1$ to n do the following:
3. Sort the values of $\vec{R}_j(i)$, $(i = 1, \ldots, k)$ in ascending order.
4. Let $N(j)$ be the number of distinct elements in $\vec{R}_j(i)$, $(i = 1, \ldots, k)$.

5. Let $\vec{X}(i) = (X_1(i), X_2(i), \ldots, X_n(i))$ be the discretized data for $\vec{R}(i)$. Assign natural numbers to $X_j(i)$ as follows: To the smallest $R_j(i)$, assign 1. To the second smallest $R_j(i)$, assign 2. To the largest $R_j(i)$, assign $N(j)$.

Example 10.2.1 Consider Table 10.1. In this case, ZTT, ALT, and ALB correspond to R_1, R_2 and R_3, respectively. Note that the examples are already sorted with respect to the value of ZTT. So, for the first example, $X_1(1) = 1$. Similarly, $X_1(2) = 2$ and $X_1(3) = 3$. Since $R_1(4) = R_1(5)$, we have $X_1(4) = X_1(5) = 4$. Also, since $R_1(6) = R_1(7) = R_1(8)$, we have $X_1(6) = X_1(7) = X_1(8) = 5$, and so on. Note that $X_1(20) = 14$, since $N(1) = 14$. In a similar way, X_2 and X_3 are derived. In this way, we have Table 10.2. ∎

Table 10.2 Blood test after discretization. © 2022 IEEE. Reprinted with permission from [2]

ID	X_1	X_2	X_3	Diagnosis
1	1	4	13	1
2	2	8	13	1
3	3	1	7	1
4	4	2	14	1
5	4	4	10	1
6	5	5	10	1
7	5	12	11	1
8	5	6	4	2
9	6	6	12	1
10	6	10	3	2
11	7	6	5	1
12	8	7	9	1
13	8	9	8	2
14	8	10	10	2
15	9	3	2	2
16	10	12	2	2
17	11	7	1	2
18	12	13	6	2
19	13	5	4	2
20	14	11	8	2

Diagnosis: 1: Negative, 2: Positive (Liver Cirrhosis)

10.2.2 Domain Reduction

Algorithm 10.2.2 *(Domain Reduction)*

1. Let $\vec{X}(i) = (X_1(i), X_2(i), \ldots, X_n(i))$ be the data for the i-th example, where $i = 1, 2, \ldots, k$, and let $Class(i)$ be the class of the i-th example.
2. For $j = 1$ to n do the following:
3. Sort the values of $\vec{X}_j(i)$, $(i = 1, 2, \ldots, k)$ in ascending order.
4. Let $\vec{Y}(i) = (Y_1(i), Y_2(i), \ldots, Y_n(i))$ be the domain-reduced data for $\vec{X}(i)$. $Y_j(i)$ is formed as follows: Let $X_j(1)$ be the smallest value, assign $Y_j(1)$ to 1. If $X_j(i + r) = X_j(i) + r$ and $Class(i + r) = Class(i)$, then $Y_j(i + r) \leftarrow Y_j(i)$. Otherwise, $Y_j(i + r) \leftarrow Y_j(i) + 1$.

Example 10.2.2 Consider Table 10.2. Since $Class(i) = 1$ for $i = 1, 2, 3, 4, 5$, we have $Y_1(i) = 1$ for $i = 1, 2, 3, 4, 5$. However, $X_i(6) = X_i(7) = X_i(8)$, but $Class(6) \neq Class(8)$, so $Y_1(6) = 2$. Also $X_1(9) = X_1(10)$, but $Class(9) \neq Class(10)$, so $Y_1(9) = 3$. In this way, we have Table 10.3. Note that the maximum values for Y_1, Y_2, and Y_3 are reduced to 6, 10, and 8, respectively.   ■

10.2.3 Multi-valued Input SOP Minimization

Algorithm 10.2.3 *(Simplification of Multi-valued Input Classification Function)*

1. Let m be the number of classes.
2. Partition the function into $(F_1, F_2, \ldots, F_m)$.
3. For each function F_i, simplify the expression by the MINI5 algorithm.[1] In this case, F_i is treated as the ON set, while $\bigcup_{j=1, j \neq i}^{m} F_j$ is treated as the OFF set.
4. Expand the input parts.
5. Merge the functions F_i $(i = 1, 2, \ldots, m)$ to form a multi-class function.

---

[1] MINI5 is an SOP minimizer for machine learning. It is similar to MINI [3], but it uses the ON and the OFF sets as inputs.

Table 10.3 Blood test after domain reduction. © 2022 IEEE. Reprinted with permission from [2]

ID	Y_1	Y_2	Y_3	Diagnosis
1	1	3	8	1
2	1	7	8	1
3	1	1	4	1
4	1	1	8	1
5	1	3	7	1
6	2	4	7	1
7	2	9	8	1
8	2	5	1	2
9	3	5	8	1
10	3	8	1	2
11	4	5	2	1
12	5	6	6	1
13	5	8	5	2
14	5	8	7	2
15	6	2	1	2
16	6	9	1	2
17	6	6	1	2
18	6	10	3	2
19	6	4	1	2
20	6	8	5	2

Diagnosis: 1: Negative, 2: Positive (Liver Cirrhosis)

Example 10.2.3 Consider the function shown in Table 10.3. In this case, Y_1 takes 6 values, Y_2 takes 10 values, and Y_3 takes 8 values. So, the size of the universe, i.e., the total number of input combinations, is $6 \times 10 \times 8 = 480$. However, F_1 and F_2 are specified by 10 combinations each. Thus, 20 combinations are specified in total. Thus, for the remaining $480 - 20 = 460$ combinations, the function values are undefined. Thus, the function is very sparse. The positional cube notation of the original data is shown below:

$$
\begin{array}{ccc}
Y_1 & Y_2 & Y_3 \\
123456 & 1234567890 & 12345678
\end{array}
$$

$$
\left[
\begin{array}{llll}
100000 & 0010000000 & 00000001 & 10 \\
100000 & 0000001000 & 00000001 & 10 \\
100000 & 1000000000 & 00010000 & 10 \\
100000 & 1000000000 & 00000001 & 10 \\
100000 & 0010000000 & 00000010 & 10 \\
010000 & 0001000000 & 00000010 & 10 \\
010000 & 0000000010 & 00000001 & 10 \\
010000 & 0000100000 & 10000000 & 01 \\
001000 & 0000100000 & 00000001 & 10 \\
001000 & 0000000100 & 10000000 & 01 \\
000100 & 0000100000 & 01000000 & 10 \\
000010 & 0000010000 & 00000100 & 10 \\
000010 & 0000000100 & 00001000 & 01 \\
000010 & 0000000100 & 00000010 & 01 \\
000001 & 0100000000 & 10000000 & 01 \\
000001 & 0000000010 & 10000000 & 01 \\
000001 & 0000010000 & 10000000 & 01 \\
000001 & 0000000001 & 00100000 & 01 \\
000001 & 0001000000 & 10000000 & 01 \\
000001 & 0000000100 & 00001000 & 01
\end{array}
\right]
$$

After SOP minimization, we have the following array:

$$
\begin{array}{ccc}
Y_1 & Y_2 & Y_3 \\
123456 & 1234567890 & 12345678
\end{array}
$$

$$
\left[
\begin{array}{llll}
111111 & 1111111011 & 01011111 & 10 \\
111111 & 1111111111 & 10101000 & 01 \\
111111 & 1100111111 & 10101010 & 01
\end{array}
\right]
$$

This array corresponds to the expressions:

$$
F_1 = \overline{Y_2^{\{8\}}} \cdot Y_3^{\{1,3\}}
$$
$$
F_2 = Y_3^{\{1,3,5\}} \vee \overline{Y_2^{\{3,4\}}} \cdot Y_3^{\{1,3,5,7\}}
$$
$$
= Y_3^{\{1,3,5\}} \vee \overline{Y_2^{\{3,4\}}} \cdot Y_3^{\{7\}}
$$

Note that
$Y_2^{\{8\}}$ corresponds to ALT $= 34$, 35, or 36;
$Y_3^{\{1\}}$ corresponds to ALB $= 3.3$, 3.5, 3.6, or 3.7;
$Y_3^{\{3\}}$ corresponds to ALB $= 3.9$;
$Y_3^{\{5\}}$ corresponds to ALB $= 4.1$;

$Y_3^{\{7\}}$ corresponds to ALB $= 4.4$;
$Y_2^{\{3,4\}}$ corresponds to ALT $= 25$ or 28.
From these, we have the following rules:

Class: Negative: (Number of examples: 10)

Rule 1   ALT : NOT $\{34, 35, 36\}$
        ALB : $\{3.8,$ or greater than $4.0\}$
        $\Longrightarrow$ Negative   (Coverage: 10)

        Class: Positive (Liver Cirrhosis):   (Number of examples: 10)

Rule 2   ALB : $\{3.3, 3.5, 3.6, 3.7, 3.9, 4.1\}$
        $\Longrightarrow$ Positive (Liver Cirrhosis)   (Coverage: 9)

Rule 3   ALT : NOT $\{25, 28\}$
        ALB : $\{4.4\}$
        $\Longrightarrow$ Positive (Liver Cirrhosis)   (Coverage: 1)

These rules are consistent, and cover all the examples in Table 10.1.   ∎

10.3   Experimental Results

We developed programs to perform the procedures presented in the previous section, and applied them to the UCI data set [4]. Table 10.4 shows experimental results. The first column shows the name of the function; the second column shows n, the original numbers of variables; the third column shows k, the number of the instances; the fourth column shows m, the number of the classes; the fifth column shows n_1, the number of the variables after SOP minimization; the sixth column shows p_1, the number of the rules after SOP minimization; the seventh column shows n_2, the number of the variables after variable minimization; and the last column shows p_2, the number of the rules after SOP minimization. Functions with * marks show that the numbers of variables were reduced by Algorithm 8.2.1, but the number of products increased.

The presented method is applicable to only consistent data sets. If there is a pair of inconsistent (conflicting) samples, one of them must be removed. The most time-consuming part of the procedure is the MVSOP minimization. After reducing the number of variables by Algorithm 8.2.1, the minimization time for MVSOPs became shorter.

The following data sets were used:

Alcohol data set contains five different types of alcohols. They are classified by QCM (quartz crystal microbalance) gas sensor. In this experiment, QCM3 was used to classify the data into five classes. Five rules were generated.

Table 10.4 Experimental results

	n	k	m	SOP Min		Var Min+ SOP Min		
				n_1	p_1	n_2	p_2	
Alcohol	11	25	5	1	5	1	5	
Caesarian	5	76	2	5	20	5	20	
Dermatology	34	356	6	17	10	6	60	*
Ecoli	7	332	6	6	31	3	48	*
Fertility	9	99	2	4	9	4	9	
Glass	9	214	6	6	14	2	37	*
Ionosphere	33	351	2	13	2	2	3	*
Iris	4	150	3	4	8	3	10	*
Shuttle	9	43500	5	9	11	4	18	*
Thyroid	21	7200	3	7	18	3	39	*
Wine	13	178	3	4	3	2	12	*

* Shows that the number of products increased after variable minimization

Caesarian contains information about caesarian section results of 80 pregnant women with the most important characteristics of delivery problems in the medical field. Four conflicting samples were removed from the original data. 20 rules were generated.

Dermatology data set contains 34 variables. The data is classified into six classes. 10 conflicting samples were removed from the original data. 10 rules were generated.

Ecoli contains data for protein localization sites. It has six classes.

Fertility data set contains semen samples provided by 100 volunteers. The samples were analyzed according to the WHO 2010 criteria. One conflicting sample was removed from the original data. 9 rules were generated.

Glass data set contains 214 samples for 6 different applications. This can be used in criminological investigation.

Ionosphere contains data for a radar system. This system consists of a phased array of 16 high-frequency antennas. The targets were free electrons in the ionosphere. "Good" radar returns are those showing evidence of some type of structure in the ionosphere. "Bad" returns are those that do not; their signals pass through the ionosphere.

Iris data set contains three classes of 50 instances each, where each class refers to a type of iris plant [5]. One class is linearly separable from the other two; the latter are NOT linearly separable from each other. The variables are

$$
\begin{array}{cccc}
Y_1 & Y_2 & Y_3 & Y_4 \\
\end{array}
$$

```
       Y1                      Y2                  Y3        Y4
12345678901234567890123 4  1234567890123456789  123456789  12345678
[ 11111111111111111111111   1111111111111111111  111111111  10000000  100 ]
  11111111111111111111111   1111111111111111111  111111110  01101000  010
  10111111111111111111111   1111111111111111111  111111001  01010100  010
  11111111111111111111111   1111111111111111111  110111111  01000100  010
  11111111111111111111111   1111111010111111111  111110110  01101010  010
  11111111111111111111111   1111111111101011111  111011101  00001011  001
  11111111111111111111111   1111111111111111111  100000111  01110011  001
[ 11111111111111111111111   1111111111111111111  111111011  00000101  001 ]
```

Fig. 10.2 Simplified expressions for iris data set

R_1 : sepal length,
R_2 : sepal width,
R_3 : petal length, and
R_4 : petal width.

Three classes are

1. Iris Setosa,
2. Iris Versicolour, and
3. Iris Virginica.

8 rules were generated.

Figure 10.2 shows the minimized array. The array shows the following expressions:

$$F_1 = Y_4^{\{1\}}$$

$$F_2 = \overline{Y_3^{\{9\}}} \cdot Y_4^{\{2,3,5\}} \vee \overline{Y_1^{\{2\}}} \cdot \overline{Y_3^{\{7,8\}}} \cdot Y_4^{\{2,4,6\}} \vee$$

$$\overline{Y_3^{\{3\}}} \cdot Y_4^{\{2,6\}} \vee \overline{Y_2^{\{8,10\}}} \cdot \overline{Y_3^{\{6,9\}}} Y_4^{\{2,3,5,7\}}$$

$$F_3 = \overline{Y_2^{\{12,14\}}} \cdot \overline{Y_3^{\{4,8\}}} \cdot Y_4^{\{5,7,8\}} \vee$$

$$Y_3^{\{1,7,8,9\}} \cdot Y_4^{\{2,3,4,7,8\}} \vee \overline{Y_3^{\{7\}}} Y_4^{\{6,8\}}$$

From these, we have the following rules:

Class: Iris Setosa: (Number of examples: 50)

Rule 1   Petal width: $\{0.1 \sim 0.6\}$
$\implies$ Iris Setosa   (Coverage: 50)

Class: Iris Versicolour: (Number of examples: 50)

Rule 2   Petal length: Less than 5.2
       Petal width: {1.0 ~ 1.4, 1.6}
       ⟹ Iris Versicolour   (Coverage: 38)

Rule 3   Sepal length: NOT {4.9}
       Petal length: NOT {5.0, 5.1}
       Petal width: {1.0 ~ 1.3, 1.5, 1.7}
       ⟹ Iris Versicolour   (Coverage: 37)

Rule 4   Petal length: NOT {4.5}
       Petal width: {1.0 ~ 1.3, 1.7}
       ⟹ Iris Versicolour   (Coverage: 28)

Rule 5   Sepal width: NOT {2.8, 3.0}
       Petal length: NOT {2.6, 2.8}
       Petal width: {1.0 ~ 1.4, 1.6, 1.7}
       ⟹ Iris Versicolour   (Coverage: 30)

   Class: Iris Virginica :   (Number of examples: 50)
Rule 6   Sepal width: NOT {3.2, 3.4}
       Petal length: NOT {4.6, 4.7, 5.1}
       Petal width: {1.6, 1.8 ~ 2.5}
       ⟹ Iris Virginica   (Coverage: 34)

Rule 7   Petal length: {1.0 ~ 1.9, 5.0 ~ 6.9}
       Petal width: {1.0 ~ 1.5, 1.8 ~ 2.5}
       ⟹ Iris Virginica   (Coverage: 43)

Rule 8   Petal length: NOT {5.0}
       Petal width: {1.7, 1.9 ~ 2.5}
       ⟹ Iris Virginica   (Coverage: 33)

These rules are consistent, and cover all the 150 examples in the data set. Note that many examples in Iris Versicolour and Iris Virginica are covered by multiple rules.

Shuttle (Satlog) contains data for five classes of shuttles.

Thyroid contains data for hypothyroid disease. There are three classes. Even after reducing the domain, one of the variables takes 237 values.

Wine set is the results of a chemical analysis of wines grown in the same region in Italy but derived from three different cultivars.

10.4   Remarks

This chapter showed a method to convert continuous variables into discrete ones.

Unlike conventional methods that use decision trees, it first reduces the domain, and then produces a sparsely defined discrete function. Then, SOPs for multi-valued input functions are simplified. The method produces consistent and a complete set of rules for a given consistent set of examples. Thus, the rules produce correct results for all the examples. For many functions, we could reduce the numbers of variables before SOP minimization. However, the reduction of variables increased the number of the products.

Related research include [6–8]. This chapter is based on [2].

10.5   Exercises

10.1   Derive the rules to classify the data in Example 10.1.1 by using J48 in WEKA system [9]. Use the default parameters of WEKA. Compare the result with the result of Fig. 10.1.

10.2   Derive the rules to classify the Iris data by using J48 in WEKA system [9]. Use the default parameters of WEKA. Compare the result with the result in this book.

References

1. Uchida O (2002) Introduction to data mining (in Japanese), Nihon Keizai Shinbun
2. Sasao T (2022) A method to generate classification rules from examples. ISMVL, May 18–22, Online, pp 176–181
3. Hong SJ, Cain RG, Ostapko DL (1974) MINI: a heuristic approach for logic minimization. IBM J Res Dev 443–458
4. https://archive.ics.uci.edu/ml/datasets.php
5. Fisher RA (1936) The use of multiple measurements in taxonomic problems. Annu Eugenics 7(Part II):179–188

6. Kahramanli S, Hacibeyoglu M, Arslan A (2011) A Boolean function approach to feature selection in consistent decision information systems. Expert Syst Appl Int J 38(7):8229–8239
7. Liu H, Hussain F, Tan CL, Dash M (2002) Discretization: an enabling technique. Data Min Knowl Discov 6:393–423
8. Sasao T, Holmgren A, Eklund P (2023) A logical method to predict outcomes after coronary artery bypass grafting. ISMVL, May 22–24
9. https://www.cs.waikato.ac.nz/ml/index.html

References on Classification Functions 11

This chapter lists papers on classification functions and index generation functions. Some publication names are abbreviated as follows:

ASPDAC for Asia-Pacific Design Automation Conference; **DAC** for ACM/IEEE Design Automation Conference; **DATE** for Design, Automation and Test in Europe; **DSD** for EUROMICRO Conference on Digital System Design, Architectures, Methods and Tools; **FPGA** for ACM International Symposium on Field Programmable Gate Arrays; **IEEE** for the Institute of Electrical and Electronics Engineers; **IEICE** for The Institute of Electronics, Information and Communication Engineers (Japan); **TX** for Transactions on X; **(E)C** for (Electronic) Computers; **CAD** for Computer-Aided Design of Integrated Circuits and Systems; **ISMVL** for IEEE International Symposium on Multiple-Valued Logic; **ICCAD** for IEEE International Conference on Computer Aided Design; **JSSC** for IEEE Journal of Solid-State Circuits; **JMVLSC** for Journal of Multiple-Valued Logic and Soft Computing; **IWLS** for International Workshop on Logic and Synthesis; **RM** for Reed-Muller Workshop; and **SASIMI** for Workshop on Synthesis And System Integration of Mixed Information technologies.

11.1 Reduction of Variables

Reduction of primitive variables for logic functions [1–8].
Heuristic method [9–16].
Iterative improvement [17–19].
Analysis by Astola's group [20–22]
Using autocorrelations [16, 23].
Using entropy functions [24].

Using non-linear transformations [25, 26, 77]
Exact minimization of compound variables [10, 27–31].
Symmetric functions [32, 33].
Extension to multi-valued inputs [34, 35].
Two-class functions [36, 37].
Non-disjoint decomposition [38].
Approximate synthesis [39].

11.2   Analysis

Number of variables [7, 8, 14, 19, 37, 40–45].
Other [46–49].

11.3   Architecture

References [50–52, 78].

11.4   Applications

Virus scanning engine [53, 54].
IP look up [55, 56].
Data mining [57].
Sensor Network [79].
Exotic particle detection in high-energy physics [58, 59].
Medical [60, 61].

11.5   Survey

References [62–67].

11.6   Miscellaneous

Set of difference vectors [15, 16, 19, 23, 45, 48, 49].
Realization of Linear Functions [68].
Books on Classification [69–76].

References

1. Brown FM (1990) Boolean reasoning: the logic of Boolean equations. Kluwer Academic Publishers, Boston
2. Borowik G, Luba T, Klempous R (2020) Comparison of algorithms for dimensionality reduction and their application to index generation functions. In: 15th international conference of system of systems engineering (SoSE), pp 283–288
3. Fujita M, Matsunaga Y (1991) Multi-level logic minimization based on minimal support and its application to the minimization of look-up table type FPGAs. In: ICCAD, pp 560–563
4. Halatsis C, Gaitanis N (1978) Irredundant normal forms and minimal dependence sets of a Boolean functions. IEEE Trans Comput C-27(11):1064–1068
5. Kambayashi Y (1979) Logic design of programmable logic arrays. IEEE Trans Comput C-28(9):609–617
6. Kuntzmann J (1965) Algèbre de Boole. Dunod, Paris. English translation: Fundamental Boolean Algebra. Blackie and Son Limited, London and Glasgow (1967)
7. Sasao T (2000) On the number of dependent variables for incompletely specified multiple-valued functions. In: ISMVL, pp 91–97, Portland, Oregon, U.S.A., May 23–25
8. Sasao T (2008) On the number of variables to represent sparse logic functions. In: ICCAD-2008, San Jose, California, USA, Nov 10–13, pp 45–51
9. Nagayama S, Sasao T, Butler JT (2016) An efficient heuristic algorithm for linear decomposition of index generation functions. In: ISMVL, May 2016, pp 96–101
10. Nagayama S, Sasao T, Butler JT (2017) A balanced decision tree based heuristic for linear decomposition of index generation functions. IEICE Trans Inf Syst E100(88):1583–1591
11. Nagayama S, Sasao T, Butler JT (2021) Improvement in the quality of solutions of a heuristic linear decomposer for index generation functions. In: ISMVL, May 25–27, pp 13–18, Virtual
12. Sasao T, Nakamura T, Matsuura M (2009) Representation of incompletely specified index generation functions using minimal number of compound variables. Patras, Greece, DSD, pp 765–772
13. Sasao T, Urano Y, Iguchi Y (2013) A heuristic method to find linear decompositions for incompletely specified index generation functions. SASIMI, Sapporo, Japan, Oct 21–22, R3-1, pp 143–148
14. Sasao T, Urano Y, Iguchi Y (2014) A lower bound on the number of variables to represent incompletely specified index generation functions. In: ISMVL, Bremen, Germany, May 19–22, pp 7–12
15. Sasao T, Urano Y, Iguchi Y (2014) A method to find linear decompositions for incompletely specified index generation functions using difference matrix. IEICE Trans Fundam Electron Commun Comput Sci E97-A(12):2427–2433
16. Sasao T (2017) A linear decomposition of index generation functions: optimization using auto-correlation functions. JMVLSC 28(1):105–127
17. Sasao T (2011) Linear transformations for variable reduction. In: RM, Tuusula, Finland, May 25–26
18. Sasao T (2015) A reduction method for the number of variables to represent index generation functions: s-Min method. In: ISMVL, May 18–20. Waterloo, Canada, pp 164–169
19. Sasao T (2019) An improved upper bound on the number of variables to represent index generation functions using linear decompositions. In: IWLS, June 21–23, Lausanne, Switzerland
20. Astola J, Astola P, Stankovic R, Tabus I (2016) An algebraic approach to reducing the number of variables of incompletely defined discrete functions. In: ISMVL, Sapporo, Japan, May 17–19, pp 107–112

21. Astola JT, Astola P, Stankovic RS, Tabus I (2017) Algebraic and combinatorial methods for reducing the number of variables of partially defined discrete functions. In: ISMVL, Novi Sad, Serbia, pp 167–172

22. Astola J, Astola P, Stankovic RS, Tabus I (2018) An Algebraic approach to reducing the number of variables of incompletely defined discrete functions. JMVLSC 31(3):239–253

23. Sasao T (2013) An application of autocorrelation functions to find linear decompositions for incompletely specified index generation functions. In: ISMVL, Toyama, Japan, May 21–24, pp 96–102

24. Simovici DA, Pletea D, Vetro R (2010) Information-theoretical mining of determining sets for partially defined functions. In: ISMVL, May 2010, pp 294–299

25. Astola H, Stankovic RS, Astola JT (2016) Index generation functions based on linear and polynomial transformations. In: ISMVL, Sapporo, Japan, May 17–19, pp 102–106

26. Astola H, Stankovic RS, Astola J (2018) Reduction of variables of index generation functions using linear and quadratic transformations. JMVLSC 31(3):255–270

27. Nagayama S, Sasao T, Butler JT (2017) An exact optimization algorithm for linear decomposition of index generation functions. In: ISMVL, May 2017, pp 161–166

28. Nagayama S, Sasao T, Butler JT (2018) An exact optimization method using ZDDs for linear decomposition of index generation function. In: ISMVL, May 16–18, Linz, Austria, pp 144–149

29. Nagayama S, Sasao T, Butler JT (2019) A dynamic programming based method for optimum linear decomposition of index generation functions. In: ISMVL, May 21–23, Fredericton, Canada, pp 144–149

30. Nagayama S, Sasao T, Butler JT (2022) A fast method for exactly optimum linear decomposition of index generation functions. JMVLSC 38:384–405

31. Sasao T, Fumishi I, Iguchi Y (2015) On exact minimization of variables for incompletely specified index generation functions using a SAT solver. In: IWLS, June 12–13, Mountain View, USA

32. Nagayama S, Sasao T, Butler JT. An exact optimization method using ZDDs for linear decomposition of symmetric index generation functions. IfCoLoG J Logics Appl 5(9):1849–1866

33. Nagayama S, Sasao T, Butler JT (2020) On optimum linear decomposition of symmetric index generation functions. In: ISMVL, Nov 9–11, Virtual, pp 144–149

34. Sasao T (2012) Multiple-valued input index generation functions: optimization by linear transformation. In: ISMVL, Victoria, Canada, May 14–16, pp 185–190

35. Sasao T (2013) Multiple-valued index generation functions: reduction of variables by linear transformation. JMVLSC 21(5–6):541–559

36. Sasao T (2019) On a minimization of variables to represent sparse multi-valued input decision functions. In: ISMVL, Fredericton, Canada, May 21–23, pp 182–187

37. Sasao T (2022) Two-level minimization for partially defined functions. In: IWLS, Online, July 18–20

38. Mazurkiewicz T (2020) Non-disjoint functional decomposition of index generation functions. In: ISMVL, Miyazaki, Japan, pp 137–142

39. Mazurkiewicz T (2022) Approximate memory-based logic synthesis of index generation functions using linear decomposition. In: ISMVL, online, pp 145–150

40. Butler JT, Sasao T (2018) Analysis of the number of variables to represent index generation functions. In: Further improvements in the Boolean domain. Cambridge Scholars Publishing, Newcastle upon Tyne, NE6 2PA, UK, pp 25–42

41. Sasao T (2008) On the number of variables to represent sparse logic functions. In: IWLS, Lake Tahoe, California, USA, June 4–6, pp 233–239

42. Sasao T (2010) On the numbers of variables to represent multi-valued incompletely specified functions. In: DSD, Lille, France, Sept 2010, pp 420–423

43. Sasao T (2014) On the average number of variables to represent incompletely specified index generation function. In: IWLS, May 30–June 1, San Francisco, CA
44. Sasao T (2018) On a memory-based realization of sparse multiple-valued functions. In: ISMVL, May 16–18, Linz, Austria, pp 50–55
45. Sasao T, Matsuura K, Iguchi Y (2019) On irreducible index generation functions. In: IWLS, June 21–23. Lausanne, Switzerland
46. Butler JT, Sasao T (2018) An exact method to enumerate decomposition charts for index generation functions. In: ISMVL, May 16–18, Linz, Austria, pp 138–143
47. Sasao T (2009) On the number of LUTs to realize sparse logic functions. In: IWLS, July 31–Aug 2
48. Sasao T (2015) On the sizes of reduced covering tables for incompletely specified index generation functions. In: RM, May 21, Waterloo, Ontario, Canada
49. Simovici DA, Zimand M, Pletea D (2012) Several remarks on index generation functions. In: ISMVL, Victoria, Canada, May 2012, pp 179–184
50. Nakahara H, Sasao T, Matsuura M (2007) A CAM emulator using look-up table cascades. In: 14th reconfigurable architectures workshop, RAW 2007, March 2007, Long Beach California, USA. CD-ROM RAW-9-paper-2
51. Sasao T (2006) A Design method of address generators using hash memories. In: IWLS, pp 102–109, Vail, Colorado, U.S.A, June 7–9
52. Sasao T, Matsuura M (2007) An implementation of an address generator using hash memories. In: DSD, Aug 27–31, Lubeck, Germany, pp 69–76
53. Nakahara H, Sasao T, Matsuura M, Kawamura Y (2009) A Parallel sieve method for a virus scanning engine. In: DSD, Patras, Greece, pp 809–816
54. Nakahara H, Sasao T, Matsuura M (2013) A virus scanning engine using an MPU and an IGU based on row-shift decomposition. IEICE Trans Inf Syst E96-D(8):1667–1675
55. Nakahara H, Sasao T, Matsuura M (2013) An architecture for IPv6 lookup using parallel index generation units. In: The 9th international symposium on applied reconfigurable computing (ARC2013), March 25–27, Los Angeles. Also, Lecture notes in computer science, vol 7806, pp 59–71
56. Nakahara H, Sasao T, Matsuura M, Iwamoto H, Terao Y (2015) A memory-based IPv6 lookup architecture using parallel index generation units. IEICE Trans Inf Syst E98-D(2):262–271
57. Wasicki D, Luba T (2022) Data analysis and mining using logical synthesis methods. In: 29th international conference on mixed design of integrated circuits and system (MIXDES), pp 165–170
58. Murovič T, Trost A (2019) Massively parallel combinational binary neural networks for edge processing. Elektrotehniski Vestnik 86(1):47–53
59. Umuroglu Y, Akhauri Y, Fraser NJ, Blott M (2020) LogicNets: co-designed neural networks and circuits for extreme-throughput applications. In: 30th international conference on field-programmable logic and applications, May 2020, pp 291–297
60. Hammer PL, Bonates TO (2006) Logical analysis of data-An overview: from combinatorial optimization to medical applications. Ann Oper Res 148(1):203–225
61. Sasao T, Holmgren A, Eklund P (2023) A logical method to predict outcomes after coronary artery bypass grafting. In: ISMVL, May 22–24
62. Sasao T (2006) Design methods for multiple-valued input address generators. In: ISMVL (invited paper), Singapore, May 17–20, pp 1–10
63. Sasao T (2011) Memory-based logic synthesis. Springer
64. Sasao T (2011) Index generation functions: recent developments. In: ISMVL, Tuusula, Finland, May 23–25, pp 1–9 (invited paper)
65. Sasao T (2011) Linear decomposition of logic functions: theory and applications. In: IWLS, San Diego, June 3–5

66. Sasao T (2014) Index generation functions: tutorial. J Multiple-Valued Logic Soft Comput 23(3–4):235–263
67. Sasao T (2017) Index generation functions: minimization algorithms. In: ISMVL, May 2017, pp 197–206 (invited paper), Novi Sad, Serbia
68. Sasao T (2018) A logic synthesis for multiple-output linear circuits. In: IWLS-2018, San Francisco, June 23–24
69. Bishop CM (2006) Pattern recognition and machine learning. Springer
70. Breiman L, Friedman JH, Stone CJ, Olshen RA (1984) Classification and regression trees. CRC Press, New York
71. Quinlan JR (1993) C4.5: programs for machine learning. Morgan Kaufmann Publishers, San Mateo, California
72. Sasao T (2019) Index generation functions. Morgan & Claypool
73. Tan PN, Steinbach M, Kumar V (2018) Introduction to data mining, 2nd edn. Pearson
74. Triantaphyllou E (2010) Data mining and knowledge discovery via logic-based methods: theory, algorithms, and applications. Springer
75. Wang L, Fu X (2005) Data mining with computational intelligence. Springer
76. Witten I, Frank E, Hall M (2016) Data mining: practical machine learning tools and techniques. Morgan Kaufmann
77. Mazurkiewicz T, Luba T (2019) Linear and non-linear decomposition of index generation functions. In: 2019 MIXDES - 26th international conference mixed design of integrated circuits and systems, pp 246–251
78. Mazurkiewicz T, Borowik G, Luba T (2018) Construction of index generation unit using probabilistic data structures. In: 2018 26th international conference on systems engineering (ICSEng), 18–20 Dec 2018, pp 1–7
79. Kokosiński Z (2021) Extraction of nonredundant information from sensor networks. In: 11th IEEE international conference on intelligent data acquisition and advanced computing systems: technology and applications, (IDAACS), pp 403–407

Conclusions

12

This chapter contains a summary of the book, and the remaining problems.

12.1   Summary of the Book

This book showed reduction methods of variables for partially defined functions. Reduction of variables not only reduces hardware, but also improves the generalization ability of classification in supervised machine learning.

Major results are

- Reduction of variables and simplification of SOPs for partially defined logic functions improves the generalization ability (Table 7.4, Example 8.3.1 and Chap. 9).
- For randomly generated two-class functions with k_0 false minterms and k_1 true minterms, the number of variables can be reduced to $\lceil \log_2(k_0 k_1) \rceil - 2$ (Theorems 5.1.1, 5.1.2).
- For sparse functions, minimization of the variable first, and then minimization of the products often reduces the PLA measure (Sect. 5.2.6).
- Any two-class function can be represented with at most $\lfloor \log_2(k_0 k_1) + 1 \rfloor$ compound variables (Theorem 6.2.2).
- Compared with neural networks, the presented method produces simpler classifiers, but accuracy is lower (Sect. 7.5).

Table 12.1 compares the SOP-based machine learning and DNN-based machine learning for classification problems. The SOP-based method finds simple rules to represent the example set. Thus, it is suitable for data mining. On the other hand, neural networks find models with high test accuracy, but they are hard to analyze, and are unsuitable to find simple rules.

© The Author(s), under exclusive license to Springer Nature Switzerland AG 2024
T. Sasao, *Classification Functions for Machine Learning and Data Mining*,
Synthesis Lectures on Digital Circuits & Systems,
https://doi.org/10.1007/978-3-031-35347-5_12

Table 12.1 Comparison of machine learning

	SOP-based method	Deep neural network
Data structure	AND-OR two-level	Multi-level threshold network
Method for learning	Reduction of variables	Backpropagation
	Reduction of products	SGD
	Reduction of literals	Batch normalization
Learning	Easy	Difficult
Learning time	Short	Long
Generalization ability	Low	High
Hardware/Power	Low	High
Required training set	Small	Large
Explainability	High	Low
Complex problems	Not applicable	Applicable
Applications	Medical diagnosis	Image classification
	Data mining	
	Packet classification	

12.2 Remaining Problems

Reducing the number of variables is a critical step in machine learning and data mining. Fortunately, efficient algorithms exist for this task. However, multiple solutions may exist with the same number of variables, requiring us to choose the one with the simplest SOP that has the fewest products.

In the case of SOPs with multi-valued inputs, some variables have higher radix than others. To optimize the solution, we need to select variables with lower radix. While these techniques improve solution quality, they also increase computation cost.

Another challenge is improving the model's generalization ability. One solution is to use an **ensemble method**, such as the one shown in Fig. 5.5.

To improve the generalization ability while maintaining simplicity and interpretability is to generate multiple simple classifiers and use voters or counters to make the final decision. This approach allows each unit to be easily interpreted compared to a single-unit realization. Moreover, it significantly reduces the required memory.

Appendix: Solutions to the Exercises

3.1 Note that x_2, x_3, and x_4 are essential variables. $\{x_1, x_2, x_3, x_4\}$ and $\{x_2, x_3, x_4, x_6\}$ can represent the function, while $\{x_2, x_3, x_4, x_5\}$ cannot. Thus, minimum sets of variables are $\{x_1, x_2, x_3, x_4\}$ and $\{x_2, x_3, x_4, x_6\}$.

3.2 When $\vec{e}_1 \notin D_r$, x_1 is not essential. So, there exists an SOP without x_1.

3.3 It is clear from the definition of an essential variable.

3.4 When $\vec{e}_1 \notin D_f$, $\vec{e}_2 \notin D_f$, and $(\vec{e}_1 \vee \vec{e}_1) \notin D_f$, both x_1 and x_2 are not essential. Next, we show that x_1 and x_2 can be removed at the same time. $\vec{e}_1 \notin D_f$ shows that $f(0, x_2, x_3, \ldots, x_n)$ and $f(1, x_2, x_3, \ldots, x_n)$ are compatible. $\vec{e}_2 \notin D_f$ shows that $f(x_1, 0, x_3, \ldots, x_n)$ and $f(x_1, 1, x_3, \ldots, x_n)$ are compatible. $(\vec{e}_1 \vee \vec{e}_2) \notin D_f$ shows that $f(0, 0, x_3, \ldots, x_n)$ and $f(1, 1, x_3, \ldots, x_n)$ are compatible, and also $f(0, 1, x_3, \ldots, x_n)$ and $f(1, 0, x_3, \ldots, x_n)$ are compatible. Thus, $f(0, 0, x_3, \ldots, x_n)$, $f(1, 1, x_3, \ldots, x_n)$, $f(0, 1, x_3, \ldots, x_n)$, and $f(1, 0, x_3, \ldots, x_n)$ are compatible. Thus, we can remove x_1 and x_2 at the same time.

3.5 To find the minimum set of variables, we have to search space of size $O(2^n)$. To find the minimum SOP, we have to search the space of $O(2^{3^n})$.

3.6 Table A.1 shows the set of difference vectors. A TAG shows the pair of vectors that generates the difference vector. In the set D_f, since $(1, 1, 1, 0) > (0, 0, 1, 0)$, $(1, 1, 1, 0)$ is deleted from D_f. The set of minimal difference vectors is $MD_f = \{(0, 1, 0, 1), (0, 0, 1, 0), (1, 0, 0, 1)\}$.
Corresponding to the set of minimal difference vectors, we have the set of clauses:

$$C(\vec{a}_1, \vec{b}_2) = y_2 \vee y_4,$$
$$C(\vec{a}_2, \vec{b}_1) = y_3, \qquad \text{and}$$
$$C(\vec{a}_2, \vec{b}_2) = y_1 \vee y_4.$$

© The Editor(s) (if applicable) and The Author(s), under exclusive license to Springer Nature Switzerland AG 2024
T. Sasao, *Classification Functions for Machine Learning and Data Mining*,
Synthesis Lectures on Digital Circuits & Systems,
https://doi.org/10.1007/978-3-031-35347-5

Table A.1 The set of difference vectors for the function in Table 3.2

x_1	x_2	x_3	x_4	TAG
1	1	1	0	$(\vec{a}_1, \vec{b}_1)$
0	1	0	1	$(\vec{a}_1, \vec{b}_2)$
0	0	1	0	$(\vec{a}_2, \vec{b}_1)$
1	0	0	1	$(\vec{a}_2, \vec{b}_2)$

The product of all the clauses is $R = (y_2 \vee y_4) y_3 (y_1 \vee y_4)$.

By converting R into an SOP, and by simplifying it, we have:

$$R = y_1 y_2 y_3 \vee y_3 y_4.$$

The minimum support set is $\{3, 4\}$. Thus, this function can be represented by x_3 and x_4.

3.7 [3] When y_4 is used, f_5 can be represented with only one variable:

$$\mathcal{F}_1 = y_4.$$

When $\{x_1, x_2, x_3, x_4\}$ are used, f_5 can be represented with four variables:

$$\mathcal{F}_2 = x_1 \bar{x}_2 \bar{x}_3 \bar{x}_4 \vee x_1 x_2 x_3 \bar{x}_4.$$

3.8 [3] For $\mathcal{F}_1$, it is clear that

$$h_{ON} \subset \mathcal{F}_1,$$

and

$$h_{OFF} \cdot \mathcal{F}_1 = 0.$$

Thus, $\mathcal{F}_1$ represents F.

F can be represented without z, since
$h_{ON}(z = 0) = 0$ and
$h_{OFF}(z = 0) = x_1 x_2 \cdots x_n y.$
$h_{ON}(z = 1) = (x_1 \oplus x_2 \oplus \cdots \oplus x_n) y \vee \bar{x}_1 \bar{x}_2 \cdots \bar{x}_n \bar{y}.$
$h_{OFF}(z = 1) = (x_1 \oplus x_2 \oplus \cdots \oplus x_n) \bar{y}$
   For $\mathcal{F}_2$, it is clear that

$$h_{ON}(z = 1) \subset \mathcal{F}_2 = \overline{y(x_1 x_2 \cdots x_n)} \vee \bar{x}_1 \bar{x}_2 \cdots \bar{x}_n,$$

and

$$h_{OFF}(z = 1) \cdot \mathcal{F}_2 = 0.$$

3.9 The number is equal to the number of *anti-chains* of n elements [1, 2, 4].

4.1 $O(kn)$.

4.2 When the function is expanded by x_1.
In this case, $Size(0) = 3$ and $Size(1) = 3$.
For the partition $x_1 = 0$: $Hist(0, 1) = 2$, $Hist(0, 2) = 1$.
For the partition $x_1 = 1$: $Hist(1, 1) = 1$, $Hist(1, 2) = 2$.
Thus, the impurity measure μ is

$$\mu = [3^2 - (2^2 + 1^2)] + [3^2 - (1^2 + 2^2)]$$
$$= (9 - 5) + (9 - 5) = 8.$$

When the function is expanded by x_2.
In this case, $Size(0) = 2$ and $Size(1) = 4$.
For the partition $x_2 = 0$: $Hist(0, 1) = 1$, $Hist(0, 2) = 1$.
For the partition $x_2 = 1$: $Hist(1, 1) = 2$, $Hist(1, 2) = 2$.
Thus, the impurity measure μ is

$$\mu = [2^2 - (1^2 + 1^2)] + [4^2 - (2^2 + 2^2)]$$
$$= (4 - 2) + (16 - 8) = 2 + 8 = 10.$$

When the function is expanded by x_3.
In this case, $Size(0) = 5$ and $Size(1) = 1$.
For the partition $x_3 = 0$: $Hist(0, 1) = 2$, $Hist(0, 2) = 3$.
For the partition $x_3 = 1$: $Hist(1, 1) = 1$, $Hist(1, 2) = 0$.
Thus, the impurity measure μ is

$$\mu = [5^2 - (2^2 + 3^2)] + [1^2 - (1^2 + 0^2)]$$
$$= (25 - 4 - 9) + (1 - 1) = 12.$$

When the function is expanded by x_4.
In this case, $Size(0) = 4$ and $Size(1) = 2$.
For the partition $x_4 = 0$: $Hist(0, 1) = 3$, $Hist(0, 2) = 1$.
For the partition $x_4 = 1$: $Hist(1, 1) = 0$, $Hist(1, 2) = 2$.
Thus, the impurity measure μ is

$$\mu = [4^2 - (3^2 + 1^2)] + [2^2 - (0^2 + 2^2)]$$
$$= (16 - 9 - 1) + (4 - 4) = 6.$$

When the function is expanded by x_5.

In this case, $Size(0) = 1$ and $Size(1) = 5$.

For the partition $x_5 = 0$: $Hist(0, 1) = 0$, $Hist(0, 2) = 1$.

For the partition $x_5 = 1$: $Hist(1, 1) = 3$, $Hist(1, 2) = 2$.

Thus, the impurity measure μ is

$$\mu = [1^2 - (0^2 + 1^2)] + [5^2 - (3^2 + 2^2)]$$
$$= (1 - 1) + (25 - 9 - 4) = 12.$$

When the function is expanded by x_6.

In this case, $Size(0) = 2$ and $Size(1) = 4$.

For the partition $x_6 = 0$: $Hist(0, 1) = 1$, $Hist(0, 2) = 1$.

For the partition $x_6 = 1$: $Hist(1, 1) = 2$, $Hist(1, 2) = 2$.

Thus, the impurity measure μ is

$$\mu = [2^2 - (1^2 + 1^2)] + [4^2 - (2^2 + 2^2)]$$
$$= (4 - 1 - 1) + (16 - 4 - 4) = 2 + 8 = 10.$$

Since x_4 yields the smallest measure, we use x_4 to expand the function.

5.1 The number of inputs to each counter is 9. For a training image representing the digit '0', the value of the counter for the digit '0' is 9. On the other hand, the values of the other counters are less than 9. Thus, any training image for digit 0 produces the correct result. This is true for other digits.

5.2 Figure A.1 shows the circuit. The number of variables for each unit is

$$p = \lceil \log_2(k_i k_j) \rceil - 2 = \lceil \log_2 10^6 \rceil - 2 = \lceil 19.93 \rceil - 2 = 18.$$

Experimental results using randomly generated function show that $p_1 = 16$ by Algorithm 3.2.1 (an exact method, 10 h), and $p_2 = 17$ by Algorithm 4.2.1 (a heuristic method, 46 milliseconds).

6.1 The number of difference vectors is at most

$$N \leq \sum_{(i<j)} k_i k_j = \binom{4}{2} \times 10^3 \times 10^3 = 6 \times 10^6 \simeq 1.4 \times 2^{22}.$$

Thus, from Theorem 6.2.2, we have $p \leq \lfloor \log_2(N + 1) \rfloor = 22$. Note that Table 6.9 shows that Algorithm 6.1.1 produced a sol with $p = 18$ compound variables.

6.2 Consider the two-class function, where $k_0 = 6000$ and $k_{Not(0)} = 54000$. Thus, by Theorem 6.2.2, the upper bound for p_0 is

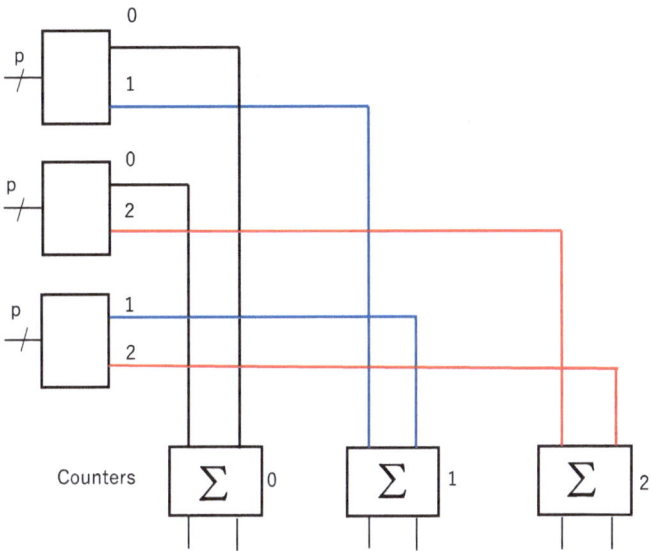

Fig. A.1 3-unit realization

Table A.2 Reduction of variables for MNIST 10-unit realization

	0	1	2	3	4	5	6	7	8	9
Primitive	30	28	31	28	28	29	29	30	30	24
Compound	23	23	23	22	23	23	22	23	21	21

$$p_0 \le \lfloor \log_2 (k_0 k_{Not(0)} + 1) \rfloor \simeq \lfloor \log_2 (3.24 \times 10^8) \rfloor = 28.$$

Thus, at most 28 compound variables are necessary.

Table A.2 shows the experimental results. When we consider the total amount of memory, the 10-unit realization requires a smaller amount of memory than the single-unit realization shown in Fig. 6.3. Note that the single-unit realization requires 25 compound variables (See Table 6.12) and 10 outputs, while the 10-unit realization require at most 23 compound variables and single output for each unit.

6.3 The number of pairs of alphabets is $\binom{26}{2} = 325$. By Lemma 6.2.1, the number of difference vectors is at most

$$N = 325 \times 780 \times 780 \simeq 1.47 \times 2^{27}.$$

Thus, by Theorem 6.2.2, the number of compound variables is at most

$$UB = \lfloor \log_2 (N + 1) \rfloor = 27.$$

When each 16-valued variable is represented by one-hot encoding, the original function can be represented by another function with $16 \times 16 = 256$ variables. With Algorithm 4.2.1 the number of primitive variables was reduced to 51. Also, with Algorithm 6.1.1, the number of compound variables was further reduced to 21.

7.1 The confusion matrix is

$$M = \begin{bmatrix} \frac{N}{2} & \frac{N}{2} \\ \frac{N}{2} & \frac{N}{2} \end{bmatrix}.$$

Thus,

$$Accuracy = Precision = Recall = 0.5,$$

while $MCC = 0.0$.

7.2 Consider the clinical example containing 9 healthy individuals and 91 sick patients [5]. Suppose that a classifier generates the following confusion matrix:

$$\begin{bmatrix} TP & FN \\ FP & TN \end{bmatrix} = \begin{bmatrix} 90 & 1 \\ 9 & 0 \end{bmatrix}.$$

In this case, the F_1 measure is $\frac{180}{180+10} = 0.947$, and shows that the performance is good. On the other hand, MCC is

$$\frac{90 \cdot 0 - 9 \cdot 1}{\sqrt{(90+9) \cdot (90+1) \cdot (0+9) \cdot (0+1)}} = -0.0316,$$

which shows the classifier is unreliable. Note that this classifier cannot identify healthy individuals.

When F_1 measure is used, the minority class should be assigned to the positive case. F_1 varies for class swapping, while MCC is invariant if the positive class is renamed negative and vice versa.

7.3 Figure A.2 shows a single-unit realization. The number of inputs to the unit is

$$p_1 = \lceil \log_2(2k \times 2k) \rceil - 2 = \lceil 2 \log_2 k \rceil.$$

Thus, the memory size is $M_1 = 2^{p_1} \simeq k^2$.

Figure A.3 shows the 2-unit realization. The number of inputs to each unit is

$$p_2 = \lceil \log_2(k \times k) \rceil - 2 = \lceil \log_2 k^2 \rceil - 2.$$

Thus, the total memory size is $M_2 = 2 \times 2^{p_2} \simeq \frac{k^2}{2}$.

A vector in the ON set in the upper (lower) unit and a vector in the OFF set in the upper (lower) unit can be distinguished correctly. However, a vector in the ON set in the upper

Fig. A.2 Single-unit realization

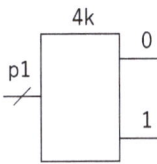

Fig. A.3 2-unit realization

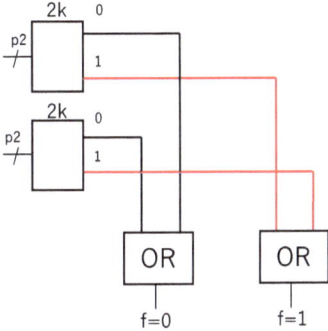

(lower) unit and a vector in the OFF set in the lower (upper) unit may not be distinguished correctly. Thus, the training accuracy can be less than 1.00.

7.4 **45-unit ×4 Realization**
Assume that each unit has at most 13 inputs and two outputs. The total memory size is

$$4 \times 2 \times \sum_{(i<j)} 2^{n(i,j)} \le 8 \times 45 \times 2^{13} \simeq 2.9 \times 10^6.$$

In addition, EXOR gates for linear circuits, counters, and the max selector [6] are necessary.

Neural Network
In this case, the number of neurons is 300, the number of inputs is $28 \times 28 = 784$. For the hidden units, the total number of bits to store the weights for the neurons is

$$8 \times 784 \times 300 \simeq 1.88 \times 10^6.$$

For the final stage, the total number of bits is

$$8 \times 300 \times 10 = 24 \times 10^3.$$

These are only for the weights for neurons. In addition, circuits for the neuron (threshold functions) are necessary.

Fig. A.4 Map for the 3-variable
function

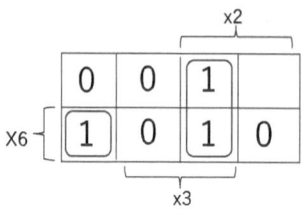

8.1 First, convert F into a disjoint SOP:

$$F = X_5^{\{2,3,4,5,7\}} \vee X_5^{\{0,1,6,8\}} X_{21}^{\{0,1,2,3,5\}} X_{22}^{\{1,5\}} \vee X_5^{\{0,1,6,8\}} X_{21}^{\{4\}} X_{22}^{\{0,2,5\}}.$$

Thus, the number of combinations covered by F is

$$|F| = 5 \times 6 \times 7 + 4 \times 5 \times 2 + 4 \times 1 \times 3 = 262.$$

Note that the size of the space of the missing variables is

$$\prod_{i=1, i \neq 5}^{20} p_i \simeq 2.58 \times 10^{11}.$$

This shows that F covers not only the poisonous mushrooms in the data set, but also other mushrooms not in the data set. However, F does not cover any edible mushrooms in the data set.

8.2 Figure A.4 shows the map of the simplified SOP. Originally, values for $k_0 + k_1 = 4 + 4 = 8$ minterms are specified. After the variable minimization, $7 \times 2^3 = 56$ minterms are specified. Thus, the function values for $56 - 8 = 48$ minterms are guessed by the variable reduction. After the SOP minimization, values for $8 \times 8 = 64$ minterms are specified. Thus, the function values for $64 - 8 = 56$ minterms are guessed.

8.3 Consider the case of 4-valued variable.

Ordinal variable. When the set of variable is $\{0, 1, 2, 3\}$, it can be represented by $\{(0, 0, 0), (1, 0, 0), (1, 1, 0), (1, 1, 1)\}$. Note that the distance between 0 and 1 is one, while the distance between 0 and 3 is three.

Nominal variable. When the values are subject of research area { mathematics, physics, chemistry, biology }, then they can be represented by $\{(1, 0, 0, 0), (0, 1, 0, 0), (0, 0, 1, 0), (0, 0, 0, 1)\}$. Note that the distance between any pair of subjects are the same.

9.1 When $n = 4$, $m = 2$.

The number of true minterms covered by the products $x_1 x_2$ and $x_3 x_4$ are 2^{n-2} each. However, the minterm covered by the product $x_1 x_2 x_3 x_4$ is counted twice. So, the number of true minterms is

$$\eta(4) = 4 + 4 - 1 = 7.$$

When $n = 6$, $m = 3$.

The number of true minterms covered by the products $x_1 x_2$, $x_3 x_4$, and $x_5 x_6$ are 2^{n-2} each. However, the minterms covered by $x_1 x_2 x_3 x_4$, $x_1 x_2 x_5 x_6$, and $x_3 x_4 x_3 x_4$, are counted twice. Also, the minterm covered by the product $x_1 x_2 x_3 x_4 x_5 x_6$ is counted three times. Thus, the number of true minterms is

$$\eta(6) = 16 + 16 + 16 - (4 + 4 + 4) + 1 = 48 - 12 + 1 = 37.$$

In general, by using the inclusion and exclusion principle, we have the formula. When $n = 12$, we have $m = 6$. Thus, $\eta(12)$ is given by

$$\binom{6}{1} 2^{12-2} - \binom{6}{2} 2^{12-4} + \binom{6}{3} 2^{12-6} - \binom{6}{4} 2^{12-8} + \binom{6}{5} 2^{12-10} - \binom{6}{6} 2^{12-12}$$
$$= 6 \cdot 2^{10} - 15 \cdot 2^8 + 20 \cdot 2^6 - 15 \cdot 2^4 + 6 \cdot 2^2 - 1 \cdot 2^0 = 3367.$$

9.2 When all the data is used for training, we have the following rules:

Class: $f = 1$ (Number of Examples: 3)

Rule 1 $x_4 = 0$

$\Longrightarrow f = 1$ (Coverage: 3/1)

Class: $f = 2$:   (Number of examples: 3)

Rule 2 $x_4 = 1$

$\Longrightarrow f = 2$ (Coverage: 2)

The number after/symbol shows incorrectly classified instances. The confusion matrix is

$$M = \begin{bmatrix} 3 & 0 \\ 1 & 2 \end{bmatrix}.$$

The number of correctly classified instances is 5, while the number of incorrectly classified instances is 1. Thus, the accuracy is 0.83.

10.1 The total number of instances is 20. The attributes are R1: ZTT, R2: ALT, and R3: ALB. When, all the data is used for training, we have the following rules:

Class: Positive (Liver Cirrhosis)   (Number of Examples: 10)

Rule 1 $ZTT \leq 12.1$
        $ALB \leq 3.7$
        $\Longrightarrow$Positive (Liver Cirrhosis)   (Coverage: 2)

Rule 2 $ZTT > 12.1$
        $\Longrightarrow$Positive (Liver Cirrhosis)   (Coverage: 9/1)

Class: Negative: (Number of examples: 10)

Rule 3 $ZTT \leq 12.1$
        $ALB > 3.7$
        $\Longrightarrow$ Negative   (Coverage: 9)

The number after/symbol shows incorrectly classified instances.
    The confusion matrix is

$$M = \begin{bmatrix} 10 & 0 \\ 1 & 9 \end{bmatrix}.$$

    The number of correctly classified instances is 19, while the number of incorrectly classified instances is 1. Thus, the accuracy is 0.95. These rules are almost equal to one in Fig. 10.1.

10.2 The total number of instances is 150. The attributes are R1: sepal length, R2: sepal width, R3: petal length, and R4: petal width. When, all the data is used for training, we have the following rules:

    Class: Iris Setosa   (Number of Examples: 50)

Rule 1 Petal width ≤ 0.6
        $\Longrightarrow$ Iris Setosa   (Coverage: 50)

Class: Iris Versicolour: (Number of examples: 50)

Rule 2 $0.6 <$ Petal width ≤ 1.7
        Petal length ≤ 4.9
        $\Longrightarrow$ Iris Versicolour   (Coverage: 48/1)

Rule 3 $1.5 <$ Petal width ≤ 1.7
$\qquad$ $4.9 <$ Petal length
$\qquad$ $\Longrightarrow$ Iris Versicolour $\qquad\qquad\qquad\qquad\qquad\qquad$ (Coverage: 3/1)

Class: Iris Virginica: $\qquad\qquad\qquad\qquad\qquad\qquad$ (Number of examples: 50)

Rule 4 $0.6 <$ Sepal width ≤ 1.5
$\qquad$ $4.9 <$ Petal length
$\qquad$ $\Longrightarrow$ Iris Virginica $\qquad\qquad\qquad\qquad\qquad\qquad$ (Coverage: 3)

Rule 5 $1.7 <$ Petal width
$\qquad$ $\Longrightarrow$ Iris Virginica $\qquad\qquad\qquad\qquad\qquad\qquad$ (Coverage: 46/1)

The number after/symbol shows incorrectly classified instances.
The confusion matrix is

$$M = \begin{bmatrix} 50 & 0 & 0 \\ 0 & 49 & 1 \\ 0 & 2 & 48 \end{bmatrix}.$$

The number of correctly classified instances is 147, while the number of incorrectly classified instances is 3. Thus, the accuracy is 0.98.

References

1. Kleitman D (1969) On Dedekind's problem: the number of monotone Boolean functions. Proc Am Math Soc 21:677–682
2. Sperner E (1928) Ein Satz uber Untermengen einer endlichen Menge. Math Z 27:544–548
3. Sasao T (2022) Two-level minimization for partially defined functions. IWLS, Online, July 18–20
4. Sasao T (2015) On the sizes of reduced covering tables for incompletely specified index generation functions. RM, May 21, Waterloo, Ontario, Canada
5. Chicco D, Jurman G (2020) The advantages of the Matthews correlation coefficient (MCC) over F1 score and accuracy in binary classification evaluation. BMC Genomics 21(6):1–13
6. Sasao T, Horikawa Y, Iguchi Y (2021) Classification functions for handwritten digit recognition. IEICE Trans Inf Syst E104-D(8):1076–1082

Index